CAMBRIDGE COUNTY GEOGRAPHIES

General Editor: F. H. H. GUILLEMARD, M.A., M.D.

BERKSHIRE

Cambridge County Geographies

BERKSHIRE

by

H. W. MONCKTON, F.L.S., F.G.S.

With Maps, Diagrams and Illustrations

Cambridge:

at the University Press

1911

CAMBRIDGE UNIVERSITY PRESS
Cambridge, New York, Melbourne, Madrid, Cape Town,
Singapore, São Paulo, Delhi, Mexico City

Cambridge University Press
The Edinburgh Building, Cambridge CB2 8RU, UK

Published in the United States of America by Cambridge University Press, New York

www.cambridge.org
Information on this title: www.cambridge.org/9781107692282

© Cambridge University Press 1911

First published 1911
First paperback edition 2013

A catalogue record for this publication is available from the British Library

ISBN 978-1-107-69228-2 Paperback

CONTENTS

CONTENTS

ILLUSTRATIONS

viii   ILLUSTRATIONS

MAPS

The illustrations on pages 7, 33, 61, 84, 88, 96, 106, are from photographs by Mr Llewellyn Treacher, of Twyford; those on pages 83 and 87 are from photographs by Mr H. A. King, of Reading; those on pages 37, 40, 46, 64, 74, 105, 158, 163 are from photographs by the author. The portraits on pages 139 and 144 are reproduced from photographs supplied by Mr Emery Walker; while the illustrations on pages 67, 69, 71, 92, 94, 97, 99, 100, 103, 110, 112, 122, 127, 129, 130, 132, 133, 153, 156, are from photographs supplied by the Homeland Association; and those on pages 2, 10, 12, 13, 16, 17, 19, 21, 22, 24, 39, 45, 58, 63, 78, 79, 98, 101, 104, 107, 115, 116, 118, 120, 121, 123, 131, 136, 141, 146, 148, 149, 151, 154, 159, 161, are from photographs supplied by Messrs F. Frith & Co., Ltd., of Reigate.

1. County and Shire. Meaning of the Words.

If we take a map of England and contrast it with a map of the United States, perhaps one of the first things we shall notice is the dissimilarity of the arbitrary divisions of land of which the countries are composed. In America the rigidly straight boundaries and rectangular shape of the majority of the States strike the eye at once; in England our wonder is rather how the boundaries have come to be so tortuous and complicated—to such a degree, indeed, that until recently many counties had outlying islands, as it were, within their neighbours' territory. We naturally infer that the conditions under which the divisions arose cannot have been the same, and that while in America these formal square blocks of land, like vast allotment gardens, were probably the creation of a central authority, and portioned off much about the same time, the divisions we find in England have no such simple origin. Such, in fact, is more or less the case. The formation of the English counties in many instances was (and is—for they have altered up to to-day) an affair of slow growth, and their origin was—as their names tell us—of very diverse nature.

Windsor Castle from the North-West

Let us turn once more to our map of England. Collectively, we call all our divisions counties, but not every one of them is accurately thus described. Some have names complete in themselves, such as Kent and Sussex, and we find these to be old English kingdoms with but little alteration either in their boundaries or their names. To others the terminal *shire* is appended, which tells us that they were *shorn* from a larger domain—*shares* of Mercia or Northumbria or some other of the great English kingdoms.

The division of England into counties or shires has often been attributed to King Alfred (A.D. 871–901), but the shire of Berks is mentioned as early as the time of Ethelbert (A.D. 860–866), and Berkshire very probably existed as a county from the days of Egbert (died 836).

The words county and shire mean practically the same thing, but the former is derived from the Latin *comitatus* through the French *comté*, the dominion of a *comes*, or Count, and the latter from the Saxon *scir* (from *sciran* to divide). The termination "shire" is generally used for Berkshire and four of the neighbouring counties, viz. Buckinghamshire, Oxfordshire, Gloucestershire, and Wiltshire. The next neighbouring county is usually called Hampshire, but in Acts of Parliament and official papers it is called the county of Southampton. For the remaining county, Surrey, the termination shire is not used : its name—Suthrege—tells us that it was "the South Kingdom."

The boundary of the county follows in great part the river Thames or its tributaries but in many places it is not distinguished from the neighbouring counties by any natural features.　On the west the chalk downs run from Wiltshire into Berkshire with no change at the boundary of the county, and on the south there is little distinction between the forest and moorland of Berkshire and of the adjoining tracts of Hampshire and Surrey.

Berkshire has thus existed as a county for about 1100 years; previously it was part of the Saxon kingdom of Wessex, which also comprised Hampshire, Wiltshire, Somerset, Dorset, Devon, and part of Cornwall.　The Saxons were called in by the Britons to assist them against the Picts and Scots (A.D. 429–449).　This was a short time after the departure of the Romans, A.D. 418, or nearly fifteen hundred years ago.　The Roman rule in our district may be taken as from A.D. 40 to 418, a period of 378 years.　We shall realise the length of their rule if we remember that 378 years ago Henry VIII was reigning in England.

When the Romans came to the district they found it occupied by a tribe of Britons named the Atrebates; and Silchester, just over our county boundary in Hampshire, was their chief town or settlement.

The written history of the district does not go further back than the Atrebates, but we find many relics of man of a much earlier date.　There are in our museums human bones found in old graves, but it is not possible to give them a date or to name the tribe or tribes to which they belonged.　There are also early gold coins without any inscription, but bearing a rude figure of a horse not unlike the celebrated white horse cut in the chalk hill above

Uffington. These coins take us back to about B.C. 200. There are also various weapons and implements of iron, bronze, and stone, found in graves or barrows or in the beds of our rivers, about which we shall say more in a subsequent chapter. All these remains belong to a period when the surface of the county, though no doubt covered to a great extent with forest, was not very different from what it is to-day. The streams and rivers followed to some extent the same courses and flowed at much the same level as now.

But there are remains of man which carry us back to a very much earlier date. In what is known as the Palaeolithic Period our rivers flowed at much higher levels than now; possibly the land has risen since that time, but however that may be, there are beds of gravel of the river Thames as much as 114 feet above the present river, and these gravels contain implements made by man. These, which are at least as old as the gravel in which we find them, are nearly all of flint, and often beautifully made. A large collection from Berkshire is in the Reading Museum.

Several animals now extinct were living at that time. The mammoth, the woolly rhinoceros, and the Irish elk roamed through the forest of Berkshire, and in all probability were hunted by Palaeolithic man.

2. General Characteristics.

Berkshire is an inland county separated from the English Channel by the full width of Hampshire. The river Thames, however, gives a waterway to the sea, and the county town, Reading, is especially well served by railways and has mainly on that account become the centre of trades of great importance. Reading biscuits and Reading seeds have a world-wide celebrity, and printing is now extensively carried on in the town.

Berkshire is, however, essentially an agricultural county, and some of the most fertile corn land in England is found in it. Until quite modern times great tracts were waste, or woodland and moorland. But these, though of no agricultural value, are for the most part very good to live in and are now being rapidly built over.

The county is divided by nature into three well-marked districts. The first of these natural divisions is formed by the Vale of White Horse and the part of the county north of it, as well as the low-lying ground between Wallingford and Steventon. The soil is clay and sand, and a few beds of limestone occur in places.

The second division is the great chalk district forming central Berkshire, with Ashbury, Wantage, and Wallingford on the north and Hungerford, Kintbury, Chieveley, Bradfield, and Tilehurst on the south. The tract included in the curve of the river Thames between Twyford and Maidenhead also belongs to the chalk district. The chalk is not always at the surface of the ground, for it is often covered by thin beds of clay or gravel, but it will always be found at a little depth below the surface in this district.

The third division comprises the forest country of the southern and south-eastern parts of the shire. Its northern boundary runs from Inkpen in the west to Maidenhead in the east, but in places tracts north of this line belong to

The Ridgeway—Uffington Castle in the distance

the third division and in other places the chalk comes to the surface south of it. The soil in the third division consists of clay and sand with no limestone. These clays and sands are very thick in the south-east of the county, but everywhere the chalk is below them if we go deep enough.

The chalk downs of the central division are dotted over with mounds and earthworks, probably for the most

part the work of man before the Roman occupation, for
it was an inhabited part of the county in the time of the
Britons.   On the other hand the Vale of White Horse
division was in those days mainly or wholly uncultivated,
but it is now the most fertile part of Berkshire.   The
south or forest division has been thinly populated up to
quite modern times, though the Roman town of Silchester
stood in the Hampshire part of this forest country.

Berkshire is almost all within the drainage area of the
river Thames and its tributaries, and the natural line of
communication between our county and the sea is by
river, Windsor being some 85 miles from the Nore.

The estuary of the Severn is less than 32 miles from
Faringdon, and there seems to have been a tolerably good
road from Berkshire to the west coast in quite early times.
Formerly a very usual line of communication between
our county and the sea was from the south coast across
the chalk downs.   Hungerford is only 35 miles from
Southampton, and the roadways across the Chalk are very
old and fairly direct.

3.   Size.   Shape.   Boundaries.

The length of Berkshire on an east and west line is
41 miles.   It may be described as a rectangle with a
somewhat square projection at the south-eastern corner.
Ashmole compares it to a lute and Fuller to a slipper.
The northern boundary is practically formed by the river
Thames, and is in consequence most irregular.   Where

the river curves in a southerly direction, the width of
the county is contracted until it is less than seven miles
at Reading. Until 1844 Three Mile Cross and the
country between that place and the Hampshire border
was an outlying part of Wiltshire, so that the width of
Berkshire at Reading was less than four miles. This
little bit of Wiltshire has however now been joined to
Berkshire.

Berkshire as it is shown upon most maps is known as
the "Geographical" or "Ancient County" of Berkshire,
and its area is 462,208 acres, that is about one-seventieth
of the area of England.

For administrative purposes the boundaries are slightly
different, and the area of Administrative Berkshire including the county borough of Reading is 462,367 acres.
By deducting from this the area under water, i.e. rivers,
ponds, lakes, etc., we arrive at the figures 459,403, which
are used as the area of Berkshire in acres for the purpose
of agricultural and other returns issued by Government.
The county of Berks for registration purposes, that is for
Parliamentary elections, etc., includes all the Administrative County and also Egham in the east, Culham and
Crowmarsh in the north-east, small bits of Oxfordshire
and Gloucestershire in the north, and the rural district of
Ramsbury in the west, giving a total area of 573,689 acres.

Berkshire was, as we have said, a part of the Saxon
kingdom of Wessex, and it has inherited from that
kingdom its northern boundary, the river Thames. It is
interesting to note that some rivers have been selected as
boundaries to a much greater extent than others. Thus

The Thames near Pangbourne

the Thames forms a county boundary for a great part of its course, whilst the river Severn flows through the middle of counties.

The Thames forms the county boundary at Old Windsor from a point a little above Magna Charta Island and separates Berkshire from Buckinghamshire, and later on from Oxfordshire, the boundary sometimes running in midstream, sometimes on one bank, and sometimes on the other bank.   Near Oxford the boundary passes for a short distance a little to the west of the river, that is on the Berks side.   The Upper Thames or Isis becomes the boundary between Berkshire and Oxfordshire, and then for a very short distance between Berkshire and Gloucestershire, until near Buscot the river Cole joins the Isis and the boundary turns in a southerly direction near to the bank of the Cole, the adjoining county being then Wiltshire.   The county boundary runs by or close to the river Cole to near Bourton, and it then crosses the chalk country with no definite marks.   At one point it crosses an old earthwork, Membury Fort, and reaches the river Kennet a little east of Chilton Foliat.   From this point to near Woodhay, a distance of some 14 miles, the boundary of the county for administrative purposes differs from the boundary of the ancient or geographical county (see page 9), indeed considerable alterations have been made in this part of the county boundary at various times. The present administrative boundary after crossing the Kennet, turns in a westerly and then in a south-easterly direction following the border of Hungerford and Inkpen parishes and runs on to a point at the south-western corner of Combe parish where Berkshire, Wiltshire, and Hampshire meet. The Berkshire boundary then runs west to Pilot Hill and then turning takes a northerly or north-easterly course until it reaches the stream Emborne which it follows for several miles until near Brimpton the stream bends sharply northwards to join the river Kennet, while the county boundary continues its easterly course

The Thames at Maidenhead

through a forest country to the Imp Stone plantation. It then makes a wide detour to the north leaving Mortimer West End and the Roman town of Silchester in Hampshire. This part of the boundary has at more than one date been subject to alteration and for a time it ran close to Silchester and is thus marked on many maps. Stratfield Mortimer is in Berkshire, and about a mile to the east of Silchester the county boundary reaches a Roman road which it follows pretty closely for a considerable distance, crossing the river Loddon at Stamford End Mill. On the east of the Loddon we come to a small tract which, until modern times, was an outlying part of Wiltshire, bounded in part by Berkshire and in part by Hampshire. It is now included in the former county, and the Berkshire boundary continues its easterly direction on or near the Roman road until it reaches the stream Whitewater

The Rivei Kennet at Hungerford

close to its junction with the Blackwater. The county boundary reaches the latter river close to a ford, no doubt a well-known place, for these fords are in most cases very old crossing-places and this one certainly goes back to Roman times and may very likely have been used in still earlier days. The boundary then turns along the Blackwater, and though it does not always follow the present course of the stream, it keeps near to it for some eight miles, until we reach the Blackwater Bridge on the London and Southampton Road. This is another ancient crossing-place, and here the counties of Berkshire, Hampshire, and Surrey meet. The Berkshire and Surrey boundary now runs in a north-easterly direction, through the grounds of the Royal Military College, Sandhurst, up a small stream to a place named Wishmoor Cross, possibly the site of a cross in former days, and evidently a well-known place, for five parishes meet there. From this point the boundary crosses the forest district of Bagshot Heath, celebrated in connection with highwaymen, and eventually reaches the Thames near Old Windsor.

In old maps it will be noticed that there are detached portions of Berkshire surrounded by Oxfordshire, and also detached portions of Wiltshire partially or wholly surrounded by Berkshire, but in modern times the county boundaries have been much modified for purposes of convenience. Thus an Act of Parliament was passed in 1844 to annex detached parts of counties to the counties in which they are situated. This Act transferred from Wiltshire to Berkshire parts of the parishes of Shinfield, Swallowfield, and Wokingham. Shilton and Little Faringdon were transferred from Berkshire to Oxfordshire, and part of Inglesham was given to Wiltshire. The boundaries of counties were still further simplified by an Act of Parliament of 1887, one of the objects of which was to arrange that no Union, Borough, Sanitary District, or Parish should be in more than one county.

4. Surface and General Features.

We have already mentioned that Berkshire may be divided into three natural divisions. The northern or Vale of White Horse district is for the most part rather low-lying ground, but there is a small range of hills along the course of the Thames or Isis from Faringdon towards Oxford. Badbury Hill, 530 feet above the sea, and Faringdon Clump, 445 feet, are quite prominent from a distance, and some of the other hills from Buckland to Wytham look imposing when seen from the river. Much of this district was to a large extent swampy and boggy ground in old days, and a part of it is still spoken of as "the moors" by the country people. Some of the village names end in "ey," suggesting that they were islands in the marsh district. Goosey and Charney are examples. A good deal of the district is stiff clay, and there is difficulty in getting a supply of good water, hence we find a number of towns and villages, like Wantage for instance, close to the chalk downs, where there are many springs.

The second or central division of Berkshire is the district of the chalk land. The downs of Berkshire are separated from the Chiltern Hills, which are the chalk hills of Oxfordshire, by the valley of the river Thames, whilst on the west the chalk downs run on into Wiltshire without any natural break. The chalk ridge rises sharply up from the Vale of White Horse, and a large part of the crest is over 700 feet above the sea. White Horse

Hill attains a height of 856 feet, and the village of
Farnborough is 712 feet above sea level. There is a
general slope of the chalk surface downwards towards
the south, so that even the high part of Lambourn
Downs is well below the 700-feet contour line, and long
and beautiful valleys run up from the Newbury district
into the chalk downs.

Crown Hill, South Ascot
(*Showing characteristics of a sandy district*)

The northern border of the chalk district is a well
defined line; not so the southern border. The chalk
gradually bends downwards underground and is covered
by sand, gravel, and clay, so that in many places we find
the upper part of the hills sandy or clayey whilst the
valleys beneath them are chalk. Thus Bussocks Camp

and Snelsmore Common near Newbury are situated upon
a ridge of gravel, sand, and clay, but the road from
Chieveley to Newbury in the valley below the camp runs
for most of the way along a chalk valley, and the chalk
extends all around, but underneath the sand, gravel, and
clay   Hence there is no definite southern boundary to

Cookham Dean

(*Showing characteristic chalk country*)

the chalk district, and there is a bit of chalk country
near Inkpen.   The projecting part of Berkshire, bounded
on the south by a line drawn from Twyford to Maidenhead and on the other sides by the river Thames, is
also mainly a chalk district.

The southern division of the county has in consequence no definite northern border, but a line drawn

from Hungerford in the west to Maidenhead in the east will have very little of chalk district to the south and very little forest country to the north, and is consequently a good practical boundary between the second and third divisions of Berkshire.

The scenery of the southern division is quite different from that of the other two divisions. The country consists to a great extent of wide and flat table-land 300 to 400 feet above the sea, in which the rivers and streams have cut valleys. There are also extensive tracts of clay land, but the clay is often concealed under a few feet of gravel.

5. Watershed. Rivers and their Courses. Lakes.

With the exception of a small tract in the south-western corner the county is wholly drained by the river Thames and its tributaries; that is to say, with a very few exceptions, every brook and stream in Berkshire is more or less directly a tributary of the Thames.

The river Thames or Isis becomes the boundary between Berkshire and Gloucestershire near Lechlade, and it flows in an easterly direction over a clay country, keeping a little to the north of the ridge of limestone hills upon which the villages of Buckland and Hinton Waldrist stand. Near Appleton the river bends to the north, curving round the outlying patch of limestone which forms Wytham Hill, and being joined by the river Evenlode. The united streams soon take a southerly

course, and a little below Oxford are joined on the north
by the Cherwell.   The river then crosses the limestone
formation near Sandford, and curves round by Radley to
Abingdon.   From Abingdon the river pursues a somewhat
serpentine course with a general south-easterly trend towards Benson, being joined on the north near Dorchester

Streatley from Goring

by the river Thame.   A little south of Benson the river,
now the Thames proper, enters upon the chalk formation, across which it flows in a southerly direction to
Streatley, and then takes a south-easterly course to Reading.
At Streatley the river valley is deep, with steep sides
separating the chalk downs of Berkshire from the chalk
hills known as the Chilterns.   The illustration above

2—2

shows the Berkshire downs in the distance and the valley of the Thames in the foreground.

At Reading the Thames is joined by the Kennet, and it is interesting to notice that the main stream adopts the direction of the tributary and flows with a north-easterly course to Wargrave, near which place the river Loddon meets it from the south, and again the direction of flow of the tributary is adopted, the Thames taking a northerly course past Henley. It is also of interest to observe that the river has turned away from the soft clays which form the ground south and east of Reading, and has cut a deep valley in the hard chalk from Wargrave onwards. Beyond Remenham the course of the river becomes easterly, and near Cookham it turns south and flows past Maidenhead to Bray.

Near Bray the Thames leaves the chalk over which it has flowed for some 40 miles and enters upon a clay country, making its way in a fairly direct line to Windsor, the one place in the district where a knob of chalk sticks up through the clay. Windsor Castle stands upon this knob of chalk. The course of the river from Bray to Windsor is on the whole south-east, and after a big curve north at Eton the course becomes more southerly, with another big curve near Old Windsor. At Runnymede House the Berkshire boundary leaves the river, which flows on to London and the sea.

The river Cole rises on the chalk not far from Ashbury, and flowing in a northerly direction joins the Upper Thames or Isis at the extreme western boundary of the county.

The river Ock rises on the chalk near Uffington, and flows down the Vale of White Horse to join the Thames at Abingdon.

The river Pang rises on the chalk not far from Compton, and flows in a southerly direction to near Bucklebury, where it turns eastward, passing through a beautiful valley by way of Stanford Dingley and

The Pang at Pangbourne

Bradfield to a point near Tidmarsh.   It then makes a sharp turn to the north and joins the Thames at Pangbourne.   This lower part of the course of the Pang is worthy of study, for there is a continuous band of river alluvium along the valley from the Thames at Pangbourne to the Kennet at Theale.   The source of the river, too, is well worthy of investigation.   In dry times it will be found in the valley near Compton, but in wet seasons it is much further up in a branch valley towards East Ilsley.

The Lambourn also rises on the chalk near the place of that name, and it flows in a south-easterly direction and joins the Kennet close to Newbury. The Pang and the Lambourn flow in chalk valleys for the whole of their course.

Pangbourne

The river Kennet rises in Wiltshire, enters Berkshire near Hungerford, and flows with an easterly course by way of Kintbury, Newbury, and Theale, finally joining the Thames close to Reading. It is a chalk river, and obtains a considerable amount of water from springs in the valley along its course.

The Emborne is not a chalk stream. It rises in the

Inkpen district and flows in an easterly direction, forming, as we have seen, the county boundary for a considerable distance.   Its course is almost parallel to that of the river Kennet, the two valleys being separated by hills or plateaux of clay, sand, and gravel.   Near Brimpton the Emborne turns sharply to the north-east, and joins the river Kennet near Sulhampstead Bannister.

The Foudry Brook rises in a clay district of Hampshire, not far from Silchester, and runs by way of Stratfield Mortimer and Grazeley to the river Kennet near Reading. It is a small stream now, but there is a good deal of alluvium along its course, showing that it was of more importance in former times.

The river Loddon rises in Hampshire and enters Berkshire at the edge of Strathfieldsaye Park, its direction being northerly.   Soon, however, it turns to the north-east and flows in a tolerably straight line to join the river Thames near Wargrave.

The Blackwater rises near Aldershot and reaches Berkshire at Blackwater Bridge, where, as we have said, the counties of Berkshire, Hampshire, and Surrey meet. From this point the river flows in a north-west or west direction and forms the Berkshire boundary for eight miles to a point near Little Ford below Farley Hill.   The Blackwater then turns into Berkshire, running in a north-westerly direction to Swallowfield, where it joins the river Loddon.

There are no natural lakes in Berkshire, though there are the deposits of a former lake in the valley of the Kennet near Newbury.

The Thames near Abingdon

There was formerly a sheet of water near Twyford named Ruscombe Lake, which had some claim to be called a natural lake, in that it was a low-lying bit of ground which was flooded owing to the absence of a good outlet. Its natural outlet was into the river Loddon, and there is a patch of alluvium extending from its site through Stanlake Park to that river. It was eventually drained by making a deep channel called the "Cut," draining a considerable area into the Thames near Bray. It has been asked why the river Thames did not follow the line of Ruscombe Lake and the Bray Cut, all soft clayey soil and low ground, instead of cutting the great and deep valley through the chalk by way of Wargrave, Henley, Great Marlow, and Maidenhead. The explanation probably is that the river Thames existed before any of these valleys, and that its course was determined by local features which have long since been destroyed by rain and streams, and by the river itself.

6. Geology and Soil.

Before giving further account of the physical geography of the county it is necessary to learn somewhat of its geology, as the physical conditions are to a large extent dependent upon geological structure.

By Geology we mean the study of the rocks, and we must at the outset explain that the term *rock* is used by the geologist without any reference to the hardness or compactness of the material to which the name is applied; thus he speaks of loose sand as a rock equally with a hard substance like granite.

Rocks are of two kinds, (1) those laid down mostly under water, (2) those due to the action of heat.

The first kind may be compared to sheets of paper one over the other. These sheets are called *beds*, and such beds are usually formed of sand (often containing pebbles), mud or clay, and limestone, or mixtures of these materials. They are laid down as flat or nearly flat sheets, but may afterwards be tilted as the result of movement of the earth's crust, just as you may tilt sheets of paper, folding them into arches and troughs, by pressing them at either end. Again, we may find the tops of the folds so produced worn away as the result of the constant action of rivers, glaciers, and sea-waves upon them, as one might cut off the tops of the folds of the paper with a pair of shears. This has happened with the ancient beds forming parts of the earth's crust, and we therefore often find them tilted, with the upper parts removed. Tilted beds are said to *dip*, the direction of dip being that in which the beds plunge *downwards*, thus the beds of an arch dip *away from* its crest, those of a trough *towards* its middle. The dip is at a low angle when the beds are nearly horizontal, and at a high angle when they approach the vertical position. The horizontal line at right angles to the direction of the dip is called the line of *strike*. Beds form strips at the surface, and the portion where they appear at the surface is called the *outcrop*. On a large scale the direction of outcrop generally corresponds with that of the strike. Beds may also be displaced along great cracks, so that one set of beds abuts against a different set at the sides of the crack, when the beds are said to be *faulted*.

The other kinds of rocks are known as igneous rocks, which have been melted under the action of heat and become solid on cooling. When in the molten state they have been poured out at the surface as the lava of volcanoes, or have been forced into other rocks and cooled in the cracks and other places of weakness. Much material is also thrown out of volcanoes as volcanic ash and dust, and is piled up on the sides of the volcano. Such ashy material may be arranged in beds, so that it partakes to some extent of the qualities of the two great rock groups.

The production of beds is of great importance to geologists, for by means of these beds we can classify the rocks according to age. If we take two sheets of paper, and lay one on the top of the other on a table, the upper one has been laid down after the other. Similarly with two beds, the upper is also the newer, and the newer will remain on the top after earth-movements, save in very exceptional cases which need not be regarded by us here, and for general purposes we may regard any bed or set of beds resting on any other in our own country as being the newer bed or set.

The movements which affect beds may occur at different times. One set of beds may be laid down flat, then thrown into folds by movement, the tops of the beds worn off, and another set of beds laid down upon the worn surface of the older beds, the edges of which will abut against the oldest of the new set of flatly deposited beds, which latter may in turn undergo disturbance and removal of their upper portions.

Again, after the formation of the beds many changes may occur in them. They may become hardened, pebble-beds being changed into conglomerates, sands into sand-stones, muds and clays into mudstones and shales, soft deposits of lime into limestone, and loose volcanic ashes into exceedingly hard rocks. They may also become cracked, and the cracks are often very regular, running in two directions at right angles one to the other. Such cracks are known as *joints*, and the joints are very important in affecting the physical geography of a district. As the result of great pressure applied sideways, the rocks may be so changed that they can be split into thin slabs, which usually, though not necessarily, split along planes standing at high angles to the horizontal. Rocks affected in this way are known as *slates*.

If we could flatten out all the beds of England, and arrange them one over the other and bore a shaft through them, we should see them on the sides of the shaft, the newest appearing at the top and the oldest at the bottom. Such a shaft would have a depth of between 50,000 and 100,000 feet. The beds are divided into three great groups called Primary or Palaeozoic, Secondary or Mesozoic, and Tertiary or Cainozoic, and at the base of the Primary rocks are the oldest rocks of Britain, which form as it were the foundation stones on which the other rocks rest, and are termed Precambrian rocks. The three

	NAMES OF SYSTEMS	SUBDIVISIONS	CHARACTERS OF ROCKS
TERTIARY	Recent Pleistocene	Metal Age Deposits Neolithic ,, Palaeolithic ,, Glacial ,,	Superficial Deposits
	Pliocene	Cromer Series Weybourne Crag Chillesford and Norwich Crags Red and Walton Crags Coralline Crag	Sands chiefly
	Miocene	Absent from Britain	
	Eocene	Fluviomarine Beds of Hampshire Bagshot Beds London Clay Oldhaven Beds, Woolwich and Reading Thanet Sands [Groups	Clays and Sands chiefly
SECONDARY	Cretaceous	Chalk Upper Greensand and Gault Lower Greensand Weald Clay Hastings Sands	Chalk at top Sandstones, Mud and Clays below
	Jurassic	Purbeck Beds Portland Beds Kimmeridge Clay Corallian Beds Oxford Clay and Kellaways Rock Cornbrash Forest Marble Great Oolite with Stonesfield Slate Inferior Oolite Lias—Upper, Middle, and Lower	Shales, Sandstones and Oolitic Limestones
	Triassic	Rhaetic Keuper Marls Keuper Sandstone Upper Bunter Sandstone Bunter Pebble Beds Lower Bunter Sandstone	Red Sandstones and Marls, Gypsum and Salt
PRIMARY	Permian	Magnesian Limestone and Sandstone Marl Slate Lower Permian Sandstone	Red Sandstones and Magnesian Limestone
	Carboniferous	Coal Measures Millstone Grit Mountain Limestone Basal Carboniferous Rocks	Sandstones, Shales and Coals at top Sandstones in middle Limestone and Shales below
	Devonian	Upper Mid } Devonian and Old Red Sand- Lower } stone	Red Sandstones, Shales, Slates and Lime- stones
	Silurian	Ludlow Beds Wenlock Beds Llandovery Beds	Sandstones, Shales and Thin Limestones
	Ordovician	Caradoc Beds Llandeilo Beds Arenig Beds	Shales, Slates, Sandstones and Thin Limestones
	Cambrian	Tremadoc Slates Lingula Flags Menevian Beds Harlech Grits and Llanberis Slates	Slates and Sandstones
	Pre-Cambrian	No definite classification yet made	Sandstones, Slates and Volcanic Rocks

great groups are divided into minor divisions known as systems.

In the preceding table (p. 29) a representation of the various great subdivisions or 'systems' of the beds which are found in the British Islands is shown. The names of the great divisions are given on the left-hand side, in the centre the chief divisions of the rocks of each system are enumerated, and on the right-hand the general characters of the rocks of each system are given.

Berkshire is now part of an island and is a long way

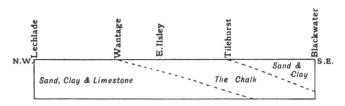

Diagram to illustrate the Geology of Berkshire

from the sea, but there have been times when the arrangement of land and sea on the globe was very different from what it is now. Our district has during some periods been part of a continent, and in others it has been overflowed by the sea.

These changes in the distribution of land and water were due to movements of the crust of the earth, and very largely to movements of compression from the sides, causing folding of the strata of which the crust of the earth is composed.

After many and great changes, at a time geologically

recent, but still long before the beginning of history in the usual sense of the word, the district now known as Berkshire rose above the sea for the last time.

Since that date deposits of clay, sand, etc., have been formed in our area, and their formation is indeed still

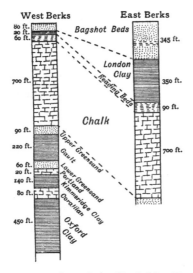

Diagram-section of the Berkshire Rocks

going on to some extent, but though these are true geological deposits they are of no great thickness, seldom as much as 20 feet. They are, however, at or near the surface of the ground, and consequently exercise considerable influence on the character of the country. We will, however, leave them out of account for the moment and

consider the deposits formed before the district finally rose
above the sea.

These deposits are usually spoken of as forming the
solid geology of the area, and the three divisions, into which
as we have said Berkshire is divided, are characterised
as follows :—

1.   In the northern part of the county, including the
Vale of White Horse, the geological strata are older than
the chalk formation.

2.   In the central part of Berkshire the chalk formation is at or near the surface of the ground.

3.   In the forest country of south and east Berkshire,
the surface is formed of geological formations newer than
the chalk, but the chalk is always to be found underground if one goes deep enough.

If we look at a sectional plan of geological strata
we shall see that none of the formations which come
to the surface in our county are of any great antiquity,
but somewhere deep down, say over a thousand feet
below us, there is a platform of much older rocks, upon
which those that come to the surface rest in an irregular
manner.   What these old rocks may be we do not know,
but probably New Red Sandstone and possibly beds of
coal may occur amongst them.

Speaking generally, we pass from older to newer geological formations as we go from the north-west towards
the south-east, and we find that the Oxford Clay is the
oldest formation which comes to the surface in Berkshire.

The Oxford Clay forms a strip of low land along the banks of the Isis from the Cole to the Cherwell near Oxford. It was originally mud deposited in a sea which extended over a great part of England. It is dark coloured, often shaley, with a little clayey limestone. A large oyster is one of its common fossils. Its thickness

Corallian Rock, Shellingford

is about 450 feet, and it is not a water-bearing formation. The Oxford Clay dips underground to the east and is covered by newer rocks, the first of which is the Corallian.

The Corallian forms a very well-marked band running across the county from the Cole to the Thames. Wytham Hill is formed of it, and Shrivenham, Coleshill, Faringdon,

Buckland, Fyfield, Appleton, and Cumnor are situated upon it. It is essentially a calcareous formation with some hard limestone beds, and has a thickness of from 50 to 80 feet. It was formed in the sea; probably a shallow sea with shoals, sand, and coral banks. Fossil corals are abundant, and many specimens of Ammonites and other marine shells are to be found. There are some good examples of these from Marcham in the Reading Museum. Supplies of good water may often be obtained from this formation. The Corallian beds are quarried for building stone and road material in many places.

The Kimmeridge Clay, which comes above the Corallian, is, like the Oxford Clay, a bed of hardened marine mud. It has now become a shaley clay, and is about 140 feet thick. It forms a narrow east and west band across the county. Much of the Vale of White Horse is on this clay, and the town of Abingdon stands upon it. It is not a water-bearing formation.

The Portland Beds. A small patch of this formation is found resting upon the Kimmeridge Clay in Berkshire. It caps the rising ground south of Shrivenham, and the village of Bourton stands upon it. Its thickness is about 20 feet.

After the deposition of the Portland rocks, which are of marine origin, there is reason to believe that our district became land and a part of a continent, but no relics of this period remain here. They were all swept away when the land sank again and the Cretaceous sea flowed over Berkshire.

The Lower Greensand—our next deposit—was formed

after a long interval, and, owing to earth movements which had taken place during that interval, it rests upon the older rocks in an irregular manner. It is a marine formation, and only occurs in patches, the largest of which extends from Uffington to near Faringdon. Its greatest thickness is about 60 feet, and it consists of sand with some ironstone and chert, pebble beds, and a calcareous sponge gravel. The sponge gravel, so-called from the number of fossil sponges it contains, is dug for garden paths and walks, and is exported to long distances. The fossil sponges in the gravel are abundant and beautifully preserved, and they seem to have lived on the spot. The ironstone was at one time worked near Faringdon. At New Lodge, in the parish of Winkfield, the Lower Greensand was reached in a boring at a depth of 1234 feet. A good supply of water was obtained, but it contains a large quantity of common salt.

The Gault, the next formation, consists of grey clay in the lower part and of a silty marl in the upper part, with a total thickness of some 220 feet. It crosses the county as a band, from one to three miles in width, from Ashbury to the Thames between Abingdon and Wallingford. It is a marine formation, and does not give a water-supply.

The Upper Greensand runs across the county as a narrow and irregular band about 90 feet thick, and consists of green sands and grey marl, with beds of stone in places. It is of marine origin, and provides a supply of excellent water, and consequently many villages stand upon or close to it. Ashbury, Childrey, Wantage, Hendred, and Harwell are examples.

3—2

The Chalk. This is far the most important geological formation in Berkshire, for it occupies a large portion of the surface of the county, and in the eastern part, when not at the surface, it is to be found underground. It is a light-coloured limestone, usually soft and earthy, but in parts very hard. Its full thickness is over 700 feet, and being a porous rock, the rain which falls on its great surface sinks in and furnishes a water-supply over its whole area whether the chalk be at the surface or underground. It was deposited in a sea which not only covered our district but spread over much of Europe. There was, however, probably land to the west which included Cornwall, parts of Wales, and of Ireland. The upper part of the Berkshire Chalk contains many layers and nodules of flint.

There is a long break in our geological record after the newest beds of the Chalk found in Berkshire had been deposited, for both the top of the Chalk and the bottom of the next series are wanting here, and in order to fill the interval we have to study rocks in other parts of England, in Belgium, and in Denmark. During this great interval in time the chalk sea retired, and much of Britain became land.

The Reading Beds repose upon a water- and weather-worn surface of chalk. They consist of clays and sands, and were deposited in the bed of a great river. Their thickness is from 70 to 90 feet, and good water may be obtained from the sands. In the lower part we find a bed of oysters, and rather higher up there is in some places a bed of leaves, known as the "Reading Leaf-Bed," a specimen of which is shown below. It will be noticed that the leaves are crowded together, and were no doubt buried in the mud of the river.

Specimen from the Reading Leaf-Bed

The Basement Bed of the London Clay comes next in order and the fossils are marine, showing that the sea was again spreading over our area. It is from 6 to 16 feet in thickness, and consists of loam and clay with green sand and pebbles. A set of shells from this bed is arranged in the Reading Museum.

The London Clay is a marine formation of very uniform character, a stiff clay, blue underground, but becoming brown near the surface, owing to the action of surface water. It contains layers of cement-stones. The thickness in the east of the county is nearly 350 feet, but the formation thins to the west, and is under 50 feet thick at Inkpen. Fossils are not uncommon, and there is a fair collection of Berkshire London Clay fossils in the Wellington College Museum. It is not a water-bearing formation. Most of Windsor Park is on London Clay, and a number of places the names of which end with "field" are upon this formation, such as Arborfield, Binfield, Burghfield, Shinfield, Swallowfield, Warfield, and Winkfield.

The Bagshot Beds, named after Bagshot Heath, consist of sand with a few beds of clay. The maximum thickness is nearly 350 feet. They are probably mainly of marine origin, but formed near the estuary of a large river. Fossils are rare in this formation in Berkshire, but a few specimens will be found in the Museums at Reading and at Wellington College. The Bagshot Beds are a water-bearing formation, but the water is not always of a satisfactory character. The scenery of the sandy Bagshot country is well shown by the view opposite.

Some indefinite time after the deposition of the Bagshot Beds considerable earth movements took place in the south of England, and Berkshire became, and has since remained, dry land.   The Bagshot Beds are consequently the last marine formation in our district, and we thus complete our account of the *solid geology* of the county.

The solid strata are, however, to a considerable extent covered with a variety of geological deposits due to rain,

Bagshot Heath Country from Bog Hill

frost, streams, and rivers.   These deposits, often termed Drift, though not marked on the majority of geological maps, have a great importance for the dwellers in our county, simply because they form the actual surface and determine the character of the soil.

Clay with Flints is a formation covering a good deal of our Chalk.   It is partly débris of the chalk formation and partly of clay beds which once rested on the Chalk.

In places it is 20 feet thick.   Some of the best timber in the county grows upon it.

Gravel covers a good deal of the surface in Berkshire. It is found both on the high ground and in the valleys. The high-level gravels are often over 10 feet thick and the valley gravels are more than 20 feet thick in several

Sarsens in Gravel, Chobham Ridges

places. Windsor, Bray, Maidenhead, Cookham, Twyford, Wokingham, Reading, Theale, Pangbourne, and Newbury stand partly or wholly upon gravel.

Alluvium, the modern deposit of the rivers, covers a good deal of ground in some places, more especially in the valley of the Kennet.

Sarsens are blocks of sandstone which are found on or

near the surface of the ground or in the beds of gravel. They were probably derived in part from the Reading Beds and in part from the Bagshot Beds. The illustration on page 40 shows three sarsen stones lying at the bottom of a thick bed of gravel in a gravel pit on Chobham Ridges. The locality is in Surrey, but not far from the Berkshire border, and similar examples occur in Berkshire.

7. Natural History.

The fertile district of the Vale of White Horse, the wide chalk downs, and the forest country with its sandy tracts covered by heather or pines, together with the river Thames and its tributaries, give us a considerable variety of soil, of climate, and of general conditions; and we consequently have a large variety of species both of animals and of plants, though being an inland county, many forms which people the coast are absent, or merely come as rare visitors. Naturally, too, the increase of population and the advance of civilisation have caused a great change in animal and plant life. Many species, once common, are no longer to be found and many new species have been introduced.

Probably the most imposing of the animals which have roamed over our district since the advent of man was the form of elephant known as the mammoth. It possessed enormous tusks and was covered with long coarse hair with an under pelage of short woolly hair

so as to be fitted for life in a cold climate. Its bones
have been found in several places in Berkshire, and teeth
from Abingdon and Reading are in the Reading Museum.

The rhinoceros once lived in Berkshire, for bones,
probably belonging to a woolly species, have been found in
a railway cutting near Chilton. Bones of the bear, wolf,
and bison have been found in the Drift deposits, and the
wild boar was hunted in Berkshire in historic times.

The badger is a harmless animal which lives a quiet
life, spending the daytime in a burrow, often in a fox
earth, and only coming out at night. It is in consequence
much more common than is generally supposed, and our
county forms no exception.

The history of the various forms of deer in Berkshire
is of considerable interest. The red deer is a native of
the county, for its remains have been found in the marsh
deposits. It lived in various parks until the Commonwealth, when most of the deer were killed. It has been
reintroduced and is now to be seen in Windsor Park,
Calcot Park, and at Hampstead Marshall. The fallow
deer lives in a more or less tame state in several parks in
the county, and it is probably an original inhabitant of
Berkshire, for it occurs as a fossil at Brentford, in Middlesex.
The roe is certainly a native, for remains have been found
in the Newbury marshes. It now lives in the woods
about Virginia Water and Sunningdale. The reindeer
has been found as a fossil at Windsor.

An imperfect skull of the musk ox was found in a
bed of gravel near Maidenhead in 1855, and is now
in the Natural History Museum at South Kensington.

It was the first discovery of the remains of this animal in Britain.

As might be expected there are no very outstanding features in Berkshire ornithology. The midland position of the county is against any long list of foreign visitors, and there are no fens or broads to tempt the special birds affecting such localities. The heron is often to be seen, and there is a heronry at Virginia Water, and others at Coley Park, Buscot, and Wytham Abbey. Woodpeckers, as might be supposed, are more especially common in the forest districts of eastern Berkshire. The carrion crow is a resident but is very local in occurrence. The hooded crow is a rather uncommon winter visitor. The peregrine falcon often visits us, but the buzzard, which used to live and breed in the county, is now but a rare visitor. The great bustard was a resident up to the end of the eighteenth century but is now no longer to be counted as a British bird. The swans which we see on the rivers and on many lakes and ponds are for the most part private property, but there are often wild birds amongst them.

Of reptiles found in Berkshire, the slow-worm, common snake, and lizard abound on the moorlands, and the first of these on the chalk ; the adder is not at all common.

Time was, and that not so very long ago, when the salmon might be caught in the Thames. In the reign of Edward III (1341), a petition was made to the King, complaining that salmon and other fish in the Thames were taken and destroyed by engines placed to catch the fry, which were then used for feeding pigs.

The King was asked to forbid the use of these engines between London and the sea, and also to decree that no salmon be taken between Gravesend and Henley bridge in winter. A book on angling published in 1815 speaking of salmon says, "some are found in the Thames which the writer believes were justly considered to be superior to any bred in other rivers."

In recent years an attempt has been made to re-introduce the salmon into the Thames, and many young salmon have been turned out in the river, but so far without any useful result.

But though the salmon has been, and again may be an inhabitant of the Thames, the brown trout is, and always has been, the fish of Berkshire. It attains a large size, and fish of from 8 to 12 lbs. are frequently caught in the Thames. There is, however, a scarcity of suitable breeding-places for trout in the river, and the stock, during recent years, has been kept up by introducing young fish, and not only brown trout but also Lochleven trout and rainbow trout have been turned into the river in great numbers. Many of the tributaries of the Thames are excellent trout streams, the Lambourn being a particularly good one.

The pike is found in the rivers and in many a lake and pond throughout Berkshire. Grayling occur in the Kennet and are occasionally caught in the Thames. The gudgeon is a well-known Thames fish ; and perch, roach, dace, barbel and minnows abound. The little ruff or pope is fairly common in the Thames, and the miller's thumb, another small fish belonging to the cooler parts of the world, is to be seen in most of our streams darting from place to place with great rapidity. The rudd, which is generally distributed through the more level part of England, is not common in Berkshire. The bream is occasionally caught in the Thames, but it is not a native and was probably introduced from Norfolk.

The Pine Plantations near Wellington College

The great variety of soil found in the river valleys, on the chalk downs, and in the forest district gives rise to much difference in the vegetation in different parts of the county. The beds of bullrush, the yellow and purple loosestrife, and the white and yellow water-lily are intimately associated with the beauty of the Thames.

The ling, the bell heather, and the cross-leaved heath cover large tracts in the eastern part of the county, and the bilberry is found in the woods of the same district. The bramble abounds in the forest parts, and of cultivated fruits we have large orchards of plums and cherries in the northern part of the county. Some rare orchids are to be found on the chalk, and in the peat districts the interesting little sundew is quite common.

Wellingtonia Avenue near Wellington College

In the chalk district the holly and beech grow well, and fine oaks are to be seen in many parts of our county. Herne's Oak, in Windsor Park, has given rise to much discussion, but there can be little doubt that the tree known by that name to Shakespeare was cut down in 1796. There are some avenues of fine elms in Windsor Park—notably the Long Walk.

Of the conifers, the yew is a native of our district and grows well on the chalk, and the so-called Scotch fir (in reality a pine), a native of Scotland, has been introduced and forms extensive woods in the sandy parts of the county. The cedar of Lebanon, various kinds of cypress, the araucaria of Chile, the cryptomeria of Japan and the Wellingtonia (*Sequoia*) of California have been introduced into the county. On the opposite page is a view of an avenue of the Wellingtonia near Wellington College.

8. Climate and Rainfall.

The climate of a country or district is, briefly, the average weather of that country or district, and it depends upon various factors, all mutually interacting, upon the latitude, the temperature, the direction and strength of the winds, the rainfall, the character of the soil, and the proximity of the district to the sea.

The differences in the climates of the world depend mainly upon latitude, but a scarcely less important factor is proximity to the sea. Along any great climatic zone there will be found variations in proportion to this proximity, the extremes being "continental" climates in the centres of continents far from the oceans, and "insular" climates in small tracts surrounded by sea. Continental climates show great differences in seasonal temperatures, the winters tending to be unusually cold and the summers unusually warm, while the climate of insular tracts is characterised by equableness and also by

greater dampness.   Great Britain possesses, by reason of
its position, a temperate insular climate, but its average
annual temperature is much higher than could be expected
from its latitude.   The prevalent south-westerly winds
cause a drift of the surface-waters of the Atlantic towards
our shores, and this warm-water current, which we know
as the Gulf Stream, is the chief cause of the mildness of
our winters.

Most of our weather comes to us from the Atlantic.
It would be impossible here within the limits of a short
chapter to discuss fully the causes which affect or control
weather changes.   It must suffice to say that the conditions
are in the main either cyclonic or anticyclonic, which
terms may be best explained, perhaps, by comparing the
air currents to a stream of water.   In a stream a chain
of eddies may often be seen fringing the more steadily-
moving central water.   Regarding the general north-
easterly moving air from the Atlantic as such a stream, a
chain of eddies may be developed in a belt parallel with
its general direction.   This belt of eddies or cyclones, as
they are termed, tends to shift its position, sometimes
passing over our islands, sometimes to the north or south
of them, and it is to this shifting that most of our weather
changes are due.   Cyclonic conditions are associated with
a greater or less amount of atmospheric disturbance ;
anticyclonic with calms.

The prevalent Atlantic winds largely affect our island
in another way, namely in its rainfall.   The air, heavily
laden with moisture from its passage over the ocean,
meets with elevated land-tracts directly it reaches our

shores—the moorland of Devon and Cornwall, the Welsh mountains, or the fells of Cumberland and Westmorland —and blowing up the rising land-surface, parts with this moisture as rain. To how great an extent this occurs is best seen by reference to the map of the annual rainfall of England on the next page, where it will at once be noticed that the heaviest fall is in the west, and that it decreases with remarkable regularity until the least fall is reached on our eastern shores. Thus in 1906, the maximum rainfall for the year occurred at Glaslyn in the Snowdon district, where 205 inches of rain fell; and the lowest was at Boyton in Suffolk, with a record of just under 20 inches. These western highlands, therefore, may not inaptly be compared to an umbrella, sheltering the country further eastward from the rain.

The above causes, then, are those mainly concerned in influencing the weather, but there are other and more local factors which often affect greatly the climate of a place, such, for example, as configuration, position, and soil. The shelter of a range of hills, a southern aspect, a sandy soil, will thus produce conditions which may differ greatly from those of a place—perhaps at no great distance—situated on a wind-swept northern slope with a cold clay soil.

Berkshire is an inland county but no part of it is as much as 75 miles from the coast. The chalk downs have a fine bracing climate, and though some of the valleys may be relaxing and some of the moorland tracts bleak, the general climate of the county is exceedingly healthy. Compared with the south coast of England

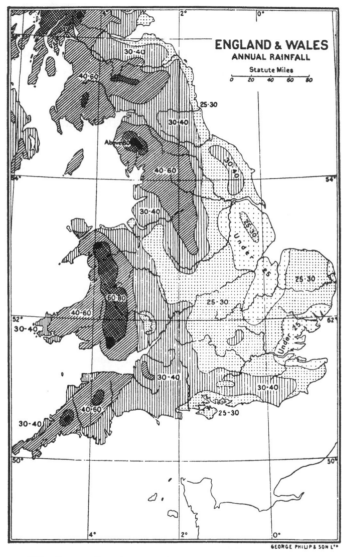

ENGLAND & WALES
ANNUAL RAINFALL
Statute Miles
0 20 40 60 80

30-40
40-60
25-30
30-40
30-40
Aberd...
40-60
30-40
Under
25
25-30
25-30
60-80
40-60
30-40
25-30
Under 25
30-40
30-40
40-60
30-40
25-30
30-40

GEORGE PHILIP & SON L'ᴰ

(The figures give the approximate annual rainfall in inches)

Berkshire is rather cooler, with somewhat less sunshine and less rain than the coast.

Temperature, it should be remarked, varies according to height above sea level, falling about 1° Fahr. for each 100 to 300 feet upwards. In a comparatively level district, like Berkshire, this is not a very serious consideration. The mean temperature for the year varies in different parts of England from about 47·3 in the north-eastern counties to about 49·6 in the south-east. The mean temperature is about 49·0 in northern Berkshire and about 47·5 in south-western Berkshire. It may be of interest to give the mean temperature for one year at places in and close around Berkshire. We take the year 1907 and the figures are as follows—Maidenhead 49·4, Wokingham 47·7, Swarraton in Hampshire 47·9, Marlborough in Wiltshire 47·4, and Oxford 48·9.

The average temperature in the month of January varies from 37·0 to 38·0 in different parts of the county, and the average temperature for July from 59·7 to 62·0.

In England bright sunshine is most prevalent on the coast and decreases inland. The annual total amount for the south and east coast from Cornwall to Norfolk is nearly 1800 hours, whilst in the northern midland counties the amount is about 1200 hours. There are no definite data available for giving the amount for Berkshire, but there are probably about 1500 hours of bright sunshine in the year.

The rainfall varies a good deal in different parts of the county. The amount is lowest in the north-east and highest in the south-west. Thus Wallingford and

Cookham have a rainfall of about 23 inches a year on an average. At Reading, which is somewhat to the south-west, the amount is nearly 24 inches a year, and on a line running through Wellington College and Yattendon the amount is nearly 25 inches. Letcombe Regis and Ashbury have a rainfall of between 25 and 26 inches. At Faringdon the figure is above 26, and in the south-western corner of the county there is a rainfall of about 29 inches a year. The average yearly rainfall for the whole of England is 31·62 inches, and for the British Isles it is 39·25 inches. Looking at the extremes of rainfall in England we find the lowest at Shoeburyness with an average of 19·7 inches for the year, whilst Seathwaite in Cumberland has an average rainfall of 133·53 inches per annum.

9. People—Race. Population.

We know little about the ancient people who made and used the flint implements which are found in the river gravels around Reading and at other places, and even when we come to the latter part of the Stone Age, though we find skeletons in the barrows or mounds upon our downs, our information about the race is exceedingly small, and this is perhaps not to be wondered at, for in no case do we find any knowledge of the art of writing in the stage of culture when only stone and no metal implements were used. Moreover, we must bear in mind that all we know about early England from written history is from the works of foreign merchants or of foreign conquerors.

The Belgae who occupied the part of Britain south of the Thames at the time of Caesar's invasion may have been partly or mostly Gauls. The tribe named the Atrebates occupied most or all of what is now Berkshire, and Silchester in Hampshire was their chief town.

During the Roman occupation the district was far from the frontier, and the inhabitants continued their peaceful village life, becoming more or less Romanised.

After the departure of the Romans the Saxons spread gradually over the country and were probably settled in Berkshire before A.D. 568. The invaders made a clean sweep of Roman civilisation, destroyed the villages and houses, and extirpated the Christian religion.

In A.D. 597 Augustine with his band of monks landed in the Isle of Thanet, and the conversion of the Saxons proceeded rapidly, and in time letters, arts, and civilisation returned to the county.

In later times Berkshire was overwhelmed by the Danes, and conquered by the Normans, but neither Danes nor Normans made anything like so great a change in the face of the country as had been effected by the Saxons, and there is even now a large Saxon element in our people, in our language, and in our manners and customs.

In early days there was a considerable population living on the chalk downs, but by degrees they moved elsewhere, and for a long time the people were mainly gathered in the valleys, especially along the banks of the

rivers Thames and Kennet. Nearly all the Berkshire
towns are situated upon one or other of those rivers.
In quite modern times there has been a great increase
of population in the eastern end of the county, large areas
of heath-land having been built over.

The population of Berkshire was steadily increasing
during the whole of the last century. In 1801 the
census gave a population of 110,752, and this had
increased in 1851 to 170,243, and in 1901 to 256,509.
That is to say the population of the county had more
than doubled in the century.

In 1901 there were 72,217 people living in the
county borough of Reading. Of the six municipal
boroughs in Berkshire Windsor had the largest population, and the others in order of numbers of inhabitants
were Maidenhead, Newbury, Abingdon, Wokingham,
and Wallingford.

Of the persons registered as inhabitants of Berkshire
in 1901, 398 were in hospital, 150 of whom were in the
Royal Berkshire Hospital at Reading; 1638 were in
Lunatic Asylums, of whom 646 were in the County
Lunatic Asylum, Cholsey; 657 in the Criminal Lunatic
Asylum at Broadmoor, and 335 in the Holloway Sana-
torium, Egham, which is in the county of Surrey, but
is included in Berkshire for registration purposes.

One man and one woman were described as over
100 years of age and they were both living in Reading.
Five men and thirteen women were described as between
95 and 100 years of age.

In the military barracks in the county there were

392 officers and 1860 non-commissioned officers and men—the 344 cadets at the Royal Military College, Sandhurst, being included amongst the officers.

The number of men engaged in the general or local government of the county was 1423. The number engaged in teaching as schoolmasters, professors, etc., was 590 men and 1712 women.

In many counties a large number of persons are described as living in ships, barges or boats, but in Berkshire the number in 1901 was only nine.

10. Agriculture.

The cultivation of the soil has probably been carried on, to some extent, since the days of the people who made the stone implements, though they doubtless chiefly concerned themselves with the chase. The early inhabitants lived partly on the chalk land and partly on the banks of the rivers. The art of cultivation no doubt spread by degrees amongst the natives, and not only the flat chalk surfaces but even the steep sides of the downs were brought into service, and they may be seen now scored with horizontal terraces in many places, partly the result of cultivation in long strips on the hill side, and partly made intentionally to assist cultivation. Terraces of this kind are found in many parts of England and are known as "linchets" or "lynchets." They form a marked feature in the landscape, near Compton Beauchamp for instance, and were at one time thought

to be old sea beaches, but this was an error ; the sea had
nothing to do with their formation. They are cultivation terraces in most cases, though in some instances they
may be, at least in part, due to landslip or to a natural
accumulation of rain wash.

During Saxon times the greater part of Berkshire
came under cultivation, and agriculture has ever since
been the main industry of our county. The Vale of
White Horse and its neighbourhood is one of the most
fertile tracts in England, and there is also some rich
pasture land on the alluvium by the rivers at Abingdon,
Purley, Newbury, Woolhampton, Theale, Reading, and
Twyford.

In the eighteenth and the early part of the nineteenth
centuries corn-growing was very profitable, and a great
deal of land was laid down in corn, some of it being
far from suited to the purpose. In later times the profit
on corn has been reduced and some of this land has been
turned to other uses or has gone out of cultivation. In
1905 the area in Berkshire devoted to corn was 98,968
acres, and in 1908 the area was 96,169 acres, a reduction
of 2799 acres. The reduction was mainly in the crop of
wheat, there was only a slight reduction in barley, whilst
there was an increase in the amount of oats. The relative
amount of wheat, barley, and oats grown in 1908 is shown
in the diagram at the end of this volume.

Berkshire is not one of the great fruit-growing counties.
In 1908 the acreage returned as orchards was 2942.

The total amount of arable land in the county in
1908 was 179,047 acres. This includes the land under

clover, sainfoin, and grasses under rotation 35,760 acres. The area of permanent grass was 175,017 acres, making a total of 354,064 acres under either crops or grass.

At the present time the production of milk is one of the most important industries of the county, the chief dairy district being the northern part and the tracts along the rivers. In 1908 the number of cattle in the county was 48,118. A cheese like single Gloucester is made in the Vale of White Horse.

The number of sheep in Berkshire was returned as 167,413. They do not belong to the breed formerly known as "Berkshire." This was a large animal with black face and black or mottled legs, which is now replaced by other kinds. The county has long been famous for its pigs, which numbered 26,171 in the year 1908.

In former days the vine was cultivated in Berkshire, and a little vineyard existed as late as the reign of George III outside Windsor Castle and to the east of Henry VIII's gateway. We also find mention of vineyards at Abingdon, Bisham, Burghfield, and Wallingford.

The number of men engaged in agriculture in Berkshire was 14,918 at the time of the last census.

11. Industries and Manufactures.

As we have said, Berkshire is essentially an agricultural county, and the cloth-making which in the days of Ashmole was so great a trade that almost the whole nation was supplied from our county, has become practi-

cally obsolete. There are however at the present day
several industries which give employment to a large
number of workers in the county. Probably the one
most definitely connected with our county town Reading
is the making of biscuits, an industry of quite modern
growth. Printing, too, is carried on at important works
at Reading belonging in many cases to London firms, and

Factory Girls leaving Work at Reading

there are also more or less active printing presses at nearly
all our towns and in country places too. Printing in
Berkshire goes back certainly to 1528, when John Scolar
set up a press in the Abbey of Abingdon and printed
a breviary, a copy of which is preserved at Emmanuel
College, Cambridge. One of the oldest of existing news-
papers is the *Reading Mercury*, started in that town in
1723.

Brewing has been carried on from the days of the monks, and no doubt plenty of good ale was brewed in the Abbeys of Abingdon and Reading. There is a record of malting mills in Wallingford Castle in 1300. At the present time there are large breweries at Reading, Windsor, and other places. Tanning is another very old industry which is still carried on with activity. The bark of the oak was formerly used to a large extent in tanning, and there has always been an abundance of oak trees in the county. Oak bark is still used to some extent. Shoe-making used to be an important cottage industry, but the introduction of machinery has carried the work to large factories elsewhere.

Newbury was at one time a great place for barge building, and boats of many kinds are now built at various places on the Thames and Kennet, indeed boat building counts amongst the more important of our active industries.

We have already mentioned cloth-making as one of the great industries of the county in former times. The chief centres were Reading, Abingdon, and Newbury. A fulling mill at Newbury is mentioned in 1205. The interesting Cloth Hall at that place, now a museum, was built by the Guild of Clothworkers of Newbury, which was incorporated in 1601, and the beautiful old house of Shaw was built by a Newbury clothier named Thomas Dolman in 1581. The most famous of the Berkshire clothiers was John Winchcombe or Smalwoode, known as Jack of Newbury (died 1520). During the seventeenth and eighteenth centuries the clothing trade declined. This was partly due to the activity of the northern clothiers and to the introduction of machinery with the resulting factory system. Still as late as 1816 there were works in Katesgrove Lane, Reading, where sail-cloth for the navy was manufactured in large quantity.

The silk industry too, once of some importance, has left the district. At the end of the sixteenth century silk-stocking making was quite an important industry at Wokingham, and many mulberry trees were planted in and near the town. Silk manufactures were also active at Reading, Newbury, Kintbury, Twyford, and other places.

Seed-growing is an important industry at Reading and employs a large number of people.

Iron and brass foundries of some importance are established at Reading and many other places, and there are large engineering works at Wantage.

There was a good bell-foundry at Wokingham in the last quarter of the fourteenth century (temp. Richard II), and several bells made there still exist. About 1495 the business was transferred to Reading, and bell-founding was carried on at that place until the beginning of the eighteenth century.

Lastly, the open country near Lambourn has long been used for training race-horses, and there are very large stables in this part of the country. The "gallops" now extend from Compton, Ilsley, and Wantage to Lambourn.

12. Minerals. Building Materials.

There is very little in the way of minerals in the rock or soil of the county. Bands of ironstone are found in the Lower Greensand formation, and it appears to have been worked near Faringdon. A group of small hollows to the east of Little Coxwell are known as Cole's Pits and were probably dug to get the iron ore.

Whitening Factory, Kintbury

Two chalybeate springs at Sunninghill were at one time quite well known.

Whiting or whitening has for a long time been manufactured at Kintbury from soft chalk which is dug there. A layer in the Reading Beds at Reading used to be dug as fuller's earth for the clothiers of that town.

Before the Norman conquest most of the buildings in the county were of wood, and of course wood has been very largely used in buildings at all times. Splendid examples of hewn timber-work may be seen in many of our churches and other buildings. There is for example some very fine old timber in the Canon's Cloisters at Windsor Castle.

Brick was a building material in the time of the Romans and its use was most probably never wholly discontinued. In Tudor times many of the buildings were of brick and timber, and picturesque brick and timber structures of various dates will be found in all parts of our county. The gallery at Christ's Hospital, Abingdon, shown on the next page, is a good example.

All the clay formations in Berkshire have been used for brick and tile making. The works at Katesgrove and other places on the banks of the Kennet at Reading are very old and certainly go back to the sixteenth century. In 1901 there were 1029 men and 35 women engaged in brick, cement, pottery, and like works in Berkshire.

The limestone rocks of the Corallian formation have been much quarried in the district between Faringdon and the river Thames near Oxford, and the stone has been used in buildings of all ages.

Chalk has also been extensively quarried for building purposes. There is a great deal of chalk in the walls of the Dean's Cloisters and also in other parts of Windsor Castle. Chalk frame-work may be seen in many church windows, at Old Windsor and Bray for instance. At Waltham St Lawrence there is a very curious example, for some flints are actually left in the chalk mullions of the east window of the north chantry. It may be of interest to mention that in the valley of the Seine in northern France chalk has been extensively used as a building stone—in some of the best buildings at Rouen for example.

Christ's Hospital, Abingdon

Flints from the chalk are much used as building-material in Berkshire; they are employed fixed in concrete to form the core of walls, as at Reading Abbey, and as facing to walls with stone corners and window-frames. Shottesbrooke Church is faced with beautifully dressed little flints. In other churches the flints are not squared but in the rough state. At St Mary's Church, Reading, there is building of a chess-board pattern, one set of squares being stone and the others formed of small dressed flints. Another example of this chequer-work is shown in the view of the church at White Waltham here given.

The hard sandstone which has been derived from the Eocene strata and is termed "sarsen" (see p. 40) is an important Berkshire building stone. There is a great deal

White Waltham Church

of it in the walls all over Windsor Castle, several of the towers and walls being faced with sarsen.

In some of the Berkshire gravel beds there is a hard irony conglomerate, and this has been used as a building material. There is a good deal in the tower of St Giles Church, Reading, and in the parish church at Wokingham.

There are many building-materials used in the county which have been brought from other districts, but this chapter only deals with things found in Berkshire itself.

Chalk was formerly used to a large extent for chalking the soil, but the practice has now almost fallen into disuse, and in consequence one sees abandoned chalk pits all over the chalk district. The reasons for giving up chalking are the increase in the cost of labour and the decrease in the value of corn crops, together with the much larger use of artificial manures. The fertility of many farms now is nevertheless due to the liming and chalking of old days, and it is to be regretted that the practice has been abandoned to so great an extent.

13. The History of Berkshire.

It has already been mentioned that Berkshire probably came into existence as a county in the time of King Egbert, who brought the long struggle between the kingdoms of the Heptarchy to a close and established the ascendancy of Wessex over much of the south of England. It is probable that there was still a population living on the chalk downs and in occupation of the old forts, and the fertile Vale of White Horse was gradually coming under cultivation. In any case there was a royal residence at Wantage, where Alfred the Great was born in 849, and a religious foundation at Abingdon. There were also at least two towns, Reading and Wallingford.

Already in the previous century the English coast had been harried by the Viking pirates, but there is no record of their having penetrated to our district. In 851 they did indeed make their way up the Thames into Surrey, but were defeated by Ethelwulf, the son of Egbert, and his son Ethelbald at Ockley. They next approached Berkshire from the south coast, and in 860 attacked and plundered Winchester, but were defeated by the united forces of Berkshire and Hampshire. Ivor the Dane is said to have reached Reading in 868, and Reading was captured and occupied by the Danes in 871.

Ethelred was at this time king and together with his b other Alfred fought the Danes near Reading, but was not successful and retreated westwards. The Danes followed and the great battle of Assandun, in which the Danes were put to flight, was fought on the chalk downs at some place to the west of Aldworth in 871. There is much doubt as to the exact site of the battle. At one time it was supposed that the White Horse was cut on the hill-side as a memorial of the victory, but it is now known that this was not so, for the horse is much older than the date of the battle. The Danes retreated to Reading, and only 14 days afterwards they got the better of the Saxons in a fight at Basing in Hampshire, and were again victorious two months later at Merton. A truce, however, followed and the Danes retired to London. All this was in the year 871, and during the same year King Ethelred died and Alfred the Great became king. How King Alfred, who ruled until 901, eventually defeated the Danes and came to terms with them is well known, and Berkshire for a time enjoyed peace.

About this time there was a royal residence at Faringdon, for it is recorded that Edward the Elder died

Statue of King Alfred, Wantage

there in 925. His son Athelstan had a mint at Wallingford, and three coins struck by him at that place are in the collection at the British Museum. The monastery at Abingdon had been destroyed by the Danes, and St Ethelwold was told by King Edred to re-establish it, but

5—2

the work was not accomplished until the reign of Edgar. Ethelred the Unready had a mint at Reading.

In 1006 the Danes again appeared in Berkshire and burnt Reading. They then advanced up the Thames to Wallingford and burnt that town. They did not, however, remain in the county, but carried their booty to the sea by way of Winchester. Both Reading and Wallingford were soon rebuilt. Edward the Confessor struck coins at both these towns, and there are specimens in the British Museum. The Confessor had a residence at Old Windsor, and the great Earl Godwin is said to have died there in a manner attributed to the judgment of God. The King gave Windsor to the Abbey of Westminster, but William the Conqueror exchanged it for some land in Essex, and built a castle on the chalk hill near the Thames where the present Windsor Castle stands. Ever since the time of the Conqueror Windsor has been a favourite residence of our Sovereigns.

In 1121 Reading Abbey was founded by King Henry I and the first Abbot was appointed in 1123. Henry added to the buildings at Windsor, and his marriage to his second wife Adelais, daughter of Godfrey Count of Louvain, took place there in 1121. There was at this time a castle at Wallingford, for it is recorded that Waleran, Earl of Mellent, was imprisoned there in 1126.

Henry I died in 1135 and was buried in Reading Abbey. On his death the peace of the county was disturbed by civil war, for the crown was claimed by Henry's nephew, Stephen of Blois, though he had sworn to support the cause of Henry's daughter Maud or

St George's Chapel, Windsor Castle

Matilda. Matilda had been married twice, and as her first husband was Henry V, the Emperor of the Holy Roman Empire, she is known as the Empress Matilda. War between Stephen and Matilda began in 1139 and spread over most of England. Windsor and Reading were held for Stephen, whilst Brian of Wallingford, a great magnate in Berkshire, took the side of Matilda. Wallingford Castle was besieged by Stephen in 1139 and again in 1145, but without success. A castle at Faringdon built by Robert Earl of Gloucester was taken and destroyed by Stephen. In 1145 Matilda gave up the contest and retired to France, but in 1152 her son Henry renewed the war and Stephen again besieged Wallingford and again unsuccessfully. He also besieged Newbury Castle, which was held by John Marshal of Hampstead Marshall. Eventually in 1153 peace was made at Wallingford— Stephen to be king for life and to be succeeded by Henry, son of Matilda. Stephen died in the next year, 1154, and Henry was crowned as King Henry II. He possessed himself of Wallingford Castle and held a Council there in 1155. Henry added to the buildings at Windsor Castle, and the lower part of the south side of the Upper Ward dates from his time.

In 1163 a duel or wager of battle was fought between Robert de Montfort and Henry of Essex on an island in the Thames below Caversham Bridge. Essex was accused of treachery or cowardice, having thrown away the standard in a battle at Coleshill. He was defeated in the duel and was allowed to join the community of Reading Abbey.

On April 19th, 1164, the ceremony of hallowing the Abbey Church at Reading was performed by Thomas à Becket, Archbishop of Canterbury, in the presence of the King. In 1175 Henry held a royal festival at Reading, and in 1185 we hear of a state ceremony at this town, when Henry received Heraclius, the Patriarch of Jerusalem.

St George's Chapel: the Interior

Henry died in 1189 and was succeeded by his son Richard I. Soon after his accession Richard left England on a crusade, having appointed the Bishops of Ely and Durham guardians of the kingdom during his absence. To his brother John he gave the government of some English districts and places, including the Honour of Wallingford. After Richard's departure a quarrel arose between the Bishop of Ely, whose name was Longchamp, and Geoffrey Archbishop of York, and Longchamp caused Geoffrey to be arrested. Prince John took the part of Geoffrey and called a Council at Reading to demand justification from Longchamp, who was summoned to meet the prince at Loddon Bridge, presumably the bridge on the Reading and Wokingham road. Longchamp did not appear, and all the participators in the arrest of the Archbishop were excommunicated in Reading church. Longchamp eventually retired to the continent, and John obtained possession of Windsor Castle, but gave it up to Queen Eleanor until Richard should come back—which he did in 1194. On Richard's death, in 1199, his brother John became King. In 1204 he obtained possession of Beckett near Shrivenham, once the property of the Earls of Evreux, and he probably lived there at times, for a mandate to the Sheriff of Oxfordshire is dated from Beckett. In 1213 John held an important ecclesiastical Council at Reading Abbey. He died in 1216 and was succeeded by his son Henry, who was in his tenth year. William Marshal, Earl of Pembroke, son of John Marshal already mentioned, was appointed Regent of the kingdom, and he held the office until his death in 1219.

In the Dean's Cloisters at Windsor may be still seen the crowned head of Henry painted during his life by William the monk of Westminster. Henry added largely to Windsor Castle, and the outer walls and towers of the Lower Ward are to a great extent his work. Disputes arose between Henry and his barons, and Berkshire was again the scene of civil war. In 1261 Parliament was summoned to meet at Windsor, and the castle was fortified by Prince Edward. It was taken in 1263 by Simon de Montfort, and the prince was captured. In time, however, he escaped and got the better of the barons.

In 1295 Berkshire sent two knights of the shire to Parliament, and Reading and Wallingford also sent representatives. In 1307 the Templars were expelled from their Preceptories at Bisham and Templeton. In the time of Edward II we hear complaints of robbers in Windsor Forest.

Edward III was born at Windsor in 1312, and his tenure of power began at a Court held at Wallingford in 1326, though his father was not deposed until the next year. King Edward wished to hold a Round Table in imitation of King Arthur, and he invited a number of knights both English and foreign to assemble at Windsor Castle in 1344. No doubt a splendid tournament took place and others followed in subsequent years. In 1347 or 1348 a garter with the motto *Hony soit qui mal y pense* was worn as a device at jousts at Windsor, and the institution of the Order of the Garter in all probability took place at Windsor in 1348, though some authorities give the date as 1349. At Christmas, 1346, the King was at Reading and a great jousting was held in his honour, and in 1359 John of Gaunt, afterwards Duke of Lancaster, was married at Reading, and there was a great pageant and a tournament in which the King and his sons took part.

During the reign of Edward III, William of Wykeham built, or re-built, the Round Tower and much of the castle at Windsor. The sword of the King is still preserved there.

In 1327 Abingdon had a little fight of its own. Some of the townspeople, assisted by the Mayor of Oxford and it is said by some scholars, attacked the Abbey and drove out the monks, part of the buildings being burnt and the muniments destroyed. In the end twelve of the attacking party were hanged and the monks restored.

Abingdon Abbey

In 1361 the Black Prince married Joan the Fair Maid of Kent. The marriage took place at Windsor, and after her husband's death Joan lived a good deal at Wallingford.

The reign of Richard II, which lasted from 1377 to 1399, was marked by constant troubles between the King with his favourites on the one hand and the nobles on the other.   In 1387 Radcot Bridge was the scene of a fight between the King's party of 5000 men under De Vere, Duke of Ireland, and Henry Earl of Derby (afterwards Henry IV).   De Vere was defeated, and only escaped by swimming down the Thames.

In 1399 Richard's inglorious reign came to an end. He was deposed in favour of Henry of Bolingbroke, son of John of Gaunt, who became King as Henry IV.

14.   The History of Berkshire (*continued*).

The reign of Henry IV lasted from 1399 to 1413. The hereditary heir to the Crown on the death of Richard II was a child, Edmund Mortimer, Earl of March, and he was detained a prisoner at Windsor Castle during the whole of Henry's reign, and only liberated by Henry V in 1413.   There was at least one fight in Berkshire during the time of Henry IV.   In 1400 an attempt was made by some of the nobles to fall on the King at Windsor, but he was warned in time, and retired to London, and when the insurgents reached Windsor, they entered the Castle without opposition, searched for the King, but found he had gone.   Meanwhile he had raised a force in London, and came to attack the insurgent nobles, who retreated, and a sharp encounter took place at Maidenhead Bridge.   The insurgents retired to Oxford and were eventually defeated.

James I, King of Scotland, was a prisoner at Windsor during most of the last ten years of his long captivity, which ended by his release in 1424. His book, *The King's Quhair*, was written at Windsor, and it was at Windsor that he fell in love with Jane Beaufort, who afterwards became his Queen.

Henry VI was born at Windsor in 1421, and became King when about nine months old. He grew up weak in mind, and during his reign all England was involved in the Wars of the Roses. Berkshire was during most of the time held by the Lancastrian party, but in 1460 Newbury was taken by the Earl of Wiltshire on behalf of the Yorkists. In the next year, 1461, the Duke of York obtained the Crown under the name of Edward IV.

Henry VI held several Parliaments at Reading, and Edward IV also visited the Abbey, and it is recorded that in 1464 he made the first public announcement of his marriage with Elizabeth Woodville at a great Council of the Peers at Reading. The marriage was not popular, and it was especially disliked by the Nevilles, the most powerful of whom, Richard Earl of Warwick, subsequently defeated Edward's forces and restored Henry VI, but Henry's renewed reign lasted only some six months, for Edward defeated Warwick, who was killed, at the battle of Barnet in 1471. Warwick and his brother the Marquis of Montagu, also killed at Barnet, were both buried at Bisham Abbey in Berkshire.

The greater part of St George's Chapel in Windsor Castle dates from the reign of King Edward IV, and he was the first of our kings to be buried there, 1483. The body of his rival Henry VI was removed to Windsor from Chertsey Abbey in 1484. The beautiful Rutland Chapel in St George's Chapel was built by Sir Thomas St Leger in memory of his wife Ann, sister of Edward IV. St Leger was beheaded by Richard III, but was buried in the chapel and a brass to himself and his wife still remains on the wall there.

After the Wars of the Roses peace reigned in Berkshire for many a long year, and the county no doubt increased in wealth and prospered generally. A considerable part of the land was in the possession of the Church, but in the days of King Henry VIII the whole of the monastic institutions were swept away.

Owing to the dissolution of the monasteries a large part of the land in Berkshire passed into the hands of the Crown. Some of it was granted to Oxford colleges and much to private persons.

In 1544 three persons, Testwood, Filmer, and Peerson were burnt at Windsor as heretics, and in 1556 Julius Palmer, Master of Reading Grammar School, John Gwin, and Thomas Askew were burnt at the sandpits near Newbury.

Elizabeth, before her accession in 1558, lived for some three years at Sir Thomas Hoby's house at Bisham; indeed she was practically a prisoner under the charge of Sir Thomas and his wife's sisters. When she came to the throne Elizabeth like her predecessors lived a good deal at Windsor, and we hear of visits by her to Reading, Englefield House (Sir F. Walsingham) and other places. It was in her days that the tragedy took place which made "Cumnor Hall" known all over the world, though its celebrity is due more to Scott's novel *Kenilworth* than to history. The real facts were, however, sufficiently tragic. Amy, the daughter of Sir John Robsart, married Lord Robert Dudley, afterwards Earl of Leicester, in 1550. Ten years later she was found dead, at Cumnor Place, which had been recently purchased by Anthony Forster the steward of Lord Robert. Foul play was suspected and it was suggested that Dudley had reasons

St George's Hall: Windsor Castle

for wishing to get rid of his wife as she stood in the way of higher ambitions. There were no "haunted towers of Cumnor Hall" for Cumnor Place was not a large house. Now only a few remains of walls are left on the site.

At the beginning of the Civil War Berkshire was generally Royalist, and the county was the scene of much fighting during the whole war, an account of which can be found in any History of England.   The Earl of Essex captured Reading after a siege in 1643, and on September 20th of the same year there was a hard-fought battle

Statue of Queen Victoria at Windsor

between Charles and Essex near Newbury.   Lord Falkland, who was on the King's side, was killed at this battle, and a granite monument to his memory stands on the high ground south of the town.

A second battle took place near Newbury on October 27th, 1644, when the Royalists occupied a position near Shaw House between the rivers Kennet and Lambourn. Earthworks, remains of this fight, may still be seen at Shaw House. Donnington Castle, near by, held out for the King until 1646, and Wallingford Castle fell into the hands of the Parliament in the same year.

On February 8th, 1649, Charles was buried in St George's Chapel at Windsor Castle.

Since the Civil War there has been only one small fight in Berkshire and that was in 1688. On December 6th of that year William of Orange reached Hungerford, and a force of 250 of his men came into conflict with 600 of James's Irish troops at Reading. Superior discipline enabled William's men to drive the Irish in confusion through the streets into the market-place where they attempted to rally, but being vigorously attacked in front, and fired upon at the same time by the inhabitants from the windows, they fled with the loss of their colours and 50 men, the conquerors only sustaining a loss of five.

There is not much to say of the history of the county since that date, though, owing to the frequent residence of the Sovereign at Windsor, many an event of the highest importance and interest has taken place there.

15. Antiquities—(*a*) Prehistoric.

We have no written records of Man as he first lived in our land long ages ago. Writing was an unknown art, and records—even if they had existed—could not have survived to come down to us. We therefore speak of this period as the Prehistoric—the time when the people of the past were unable themselves to record their story. Yet, though these sources of information are closed to us, we are able from the relics they have left behind them— the implements and weapons that they used, the bones of the animals they fed upon, the structures they erected— to form a fairly clear idea of these early peoples.

But this Prehistoric period, vast in its extent, has for convenience sake been further subdivided. At first the metals were unknown, or at least unused, and this period is spoken of as the Stone Age, for it was of flints and other stones that weapons and domestic implements were mainly fashioned. Later, man learnt how to get the easily-worked ores of tin and copper from the rocks and by their admixture to form bronze. From this, beautiful weapons and other articles were made, and from the time of the discovery we date what is known as the Bronze Age. Doubtless the ores of iron had long been known, but how to smelt them was another matter. At length the method was discovered, and mankind was in possession of hard metal implements having great advantages for all purposes over those previously employed. Thus the Iron Age began, and the early inhabitants of Britain had arrived at this stage of civilisation when the Romans came to our land.

We may now turn to a consideration of these various epochs in their order. Firstly the Stone Age. This, though a convenient term as covering all the period before the advent of the Metal Ages, is too indefinite both as to time and race, and hence it is usual to speak of the Palaeolithic or Old Stone Age, and the Neolithic or New Stone Age. The people of these two Ages were very distinct, and most authorities hold that—at all events in our land—a vast gap of time separated them, though no such gap occurred between the later Ages. Palaeolithic man, from various causes, ceased to inhabit what we now call Britain, and when the country was re-peopled it was by Neolithic man. Palaeolithic man lived in the days when the mammoth, reindeer, and hyaena roamed over our country; made leaf-shaped roughly-flaked flint weapons which were never ground or polished; cultivated no plants and tamed no animals; and built no monuments, graves, or houses. Neolithic man, on the other hand, learnt how to grind and polish his implements; was both a farmer and a breeder of stock; had many industries; and built megalithic monuments, houses, and graves—the remains of which survive to the present day.

The earliest signs of the existence of man in Berkshire are, as we have said, the implements of stone, mostly flint, found in the gravels; and the implements of the Palaeolithic Period take us back to a very old time, so old that the surface features of our district were then quite different from what we see now.

There is a fine series of Palaeolithic implements in the Reading Museum, and most of them have been found in gravel-pits near the river Thames in the Reading and Twyford district, or in the Cookham and Maidenhead district. The implements occur in the gravel in such a way as to prove that they were brought into the position in which we find them at the same time and in the same manner as the other stones in the gravel, and the men who made them consequently lived at or before the date of the making of the beds of gravel.   All the gravels in question were made by our rivers, and as the places where we find the implements are in some cases from 85 to 114

Wayland Smith's Cave

feet above the present level of the river, we infer that the valley has been deepened as much as from 85 to 114 feet since the time when the men who made the implements lived.

We now come to the Neolithic Period when, as we have seen, man was a much more civilised person than

6—2

the earlier man is believed to have been. Some of his burial mounds still remain, and being oval in plan are known as long barrows. Wayland Smith's Cave, a mile to the east of Ashbury (p. 83), is composed of some 32 stones, the remains of a long barrow of Neolithic times.

Flint Implements of the Neolithic Period found in Berkshire

Neolithic implements are of stone, but in many cases they are unlike the older implements in being of polished stone. In the Reading Museum there is a fine polished flint chisel from Englefield, and also polished axes from

Broadmoor, from Pangbourne, and from the beds of the Thames and Kennet. In the British Museum there is a beautiful dagger of flint from a barrow on Lambourn Down. Pretty little arrow-heads have been found at many places on the downs and in the Wallingford district.

There was in Berkshire a long interval between the Palaeolithic and the Neolithic Periods, but so far as we know there was no such break between the Neolithic Period and the Bronze Age. All we can say is that there was a time when the inhabitants of our district began to use implements of copper, or of copper alloyed with tin, i.e. bronze, for some purposes, but they still continued to use implements of stone, and it is not always possible to say whether a stone implement belongs to the Neolithic Period, to the Bronze Age, or to an even later date.

Many remains of the Bronze Age have been found in burial-mounds or barrows, and the barrows of this period are circular, with a diameter of fifty to one hundred feet, and hence termed round barrows. Many pieces of sepulchral pottery of this age from Berkshire will be found both in the British Museum and the Reading Museum. A considerable number of bronze implements were found in one place at Yattendon, and another hoard of them was discovered at Wallingford. A great many bronze swords, daggers, and spear-heads have been found in the river Thames, and are to be seen in the Museums.

A cemetery of this period was found at Sulham, and many earthenware urns from it are in the Reading

Museum. There are also in the Museum some urns from Neolithic barrows at Sunningdale.

The extensive deposits of peat at and around Newbury show that it was a marsh and lake district until historical time, and remains of pile dwellings have been found in the market-place, in Bartholomew Street, and in Cheap Street. Their date cannot be fixed with certainty, but they are almost certainly prehistoric in age.

The substitution of iron for bronze indicates a considerable advance in knowledge, for, except in meteorites, pure iron is not found in nature, and no small skill is required to separate the metal from the earth or rock in which it occurs. There is, however, no definite division between the Bronze and the Iron Age, for implements and ornaments of both bronze and stone continued to be used. Nor is there any definite end to the Iron Age: it passes onwards into the period of written history.

A number of bones and various objects found in a grave on Hagbourne Hill seemed to show that a man, a horse, and possibly also a chariot had been buried there.

Ancient British coins have been found at Brightwell, Newbury, Wallingford, and at other places in Berkshire. Many of them bear on one side a rude representation of a horse, probably an imitation of the horse on the gold stater of Philip II of Macedon, who became king in B.C. 359. These gold coins, known as Philips, were current in Greece and in the East for a long period, and have been occasionally found in circulation even in modern times. The White Horse, which is cut in the turf on the chalk hill above Uffington, bears a considerable resemblance to the horse on the British coins, and may very probably be of the same date.

There are a great number of mounds and earthworks scattered over Berkshire, and it is exceedingly difficult to assign to them their proper dates.   We have already mentioned Wayland Smith's Cave as the remains of a

The White Horse

long barrow of the Neolithic Period, and we have also referred to the round barrows of the Bronze Age.   Some of the fortifications may date from these early times but many are probably of later date.   It was for a long time needful to provide defence for the dwellings, not only against men, but also against wild animals, and the earthworks were no doubt used over and over again by successive peoples.

As we have said, the chalk district was at one time the most populous part of the county, and we consequently find the downs dotted over with mounds and earthworks of very ancient date. Perhaps the best known of these is the fine earthwork named Uffington Castle on White Horse Hill (see p. 7). Alfred's Castle is a circular earthwork close to Ashdown Park and three miles south-west of Uffington Castle. Letcombe Castle is another fair-sized work on

Blewburton Hill, near Blewbury

the Ridge Way, rather more than five miles east of White Horse Hill. There is a large earthwork called Danish Camp on Blewburton Hill to the south of Didcot.

There are a few old earthworks in the Vale of White Horse district. One crowns Badbury Hill near Faringdon. Cherbury Camp is a large oval work on low ground near Buckland. Sinodun Hill to the north of Wallingford has evidently been fortified in early times, and Wallingford itself has the remains of an old and extensive earthwork round the town.

Passing to the Forest District we find many mounds and banks on the heaths, and there is one very fine earthwork known as Caesar's Camp near Easthampstead. It was very likely used by the Romans, but is almost certainly of still older date. Finally it is highly probable that Windsor Castle stands on the site of an old fort.

16.   Antiquities—(*b*) **Roman and Saxon.**

The Reading Museum contains one of the finest Anglo-Roman collections in England. It is the result of careful and systematic excavation, carried on for a series of years, on the site of the town of Silchester, and the collection is of the greatest interest to us as illustrating the life in an English country town in the days of the Romans. The locality is however in Hampshire, the Berkshire boundary making a detour so as to leave it in the neighbouring county.

According to the ordnance map, Speen House near Newbury was the site of the Roman *Spinae*, but no Roman remains have been found there, though there is evidence of a settlement of some importance at Newbury itself.

The foundations of houses of the Roman period have been found at several places in Berkshire; thus at Frilford near Marcham the remains of a small Romano-British house were found ; and near by, in Frilford Field, a cemetery of the same period, which had subsequently been used by the Anglo-Saxons. Remains of a house with tessellated pavements were found on the Great Western Railway at Basildon, and other remains of Romano-British buildings have been discovered near Maidenhead and Waltham St Lawrence.

The words "Roman Villa" will be found marked on the ordnance map at two places to the south of Hampstead Norris, and remains of buildings have been discovered near Letcombe Regis, and at other places. The earthworks on Lowbury Hill to the west of Streatley are usually believed to be a Roman camp, and it is probable that the Roman soldiers occupied many of the old British forts at one time or another.

Roman coins and pottery of the Romano-British period have been found almost all over the county, though they may be said to be most common along the valley of the Thames and least so near Faringdon. In the Reading Museum there are a good many objects of Roman date which were found in Reading itself. Specimens are exhibited from two small hoards of coins dating from the Emperor Valentinian A.D. 364 to the Emperor Honorius A.D. 423. The coins are in very good preservation and were probably hidden when the Roman soldiers departed from England.

There are signs of Roman settlements along the Devil's Highway, the road from Silchester to London. Thus there was evidently a Romano-British village at Wickham Bushes close to Caesar's Camp on Easthampstead Plain. A collection from this locality exists at Wellington College.

A number of objects of the Anglo-Saxon period found in Berkshire will be seen in the Anglo-Saxon room at the British Museum. There is a very fine sword-blade from Ashdown, and a variety of objects—shield-bosses, knives, etc.—from Long Wittenham, where a Saxon burial-place has been explored. In some cases the body had been burnt, whilst in others the skeletons remained, and were found to be of a large-sized and robust race. Another Anglo-Saxon cemetery was discovered at Arne Hill near Lockinge, and a number of Anglo-Saxon interments in the Lambourn valley near East Shefford. Two burial-places of this period have been found at Reading. One contained spear-heads, knives, and bronze ornaments, and was probably of pagan date, whilst the other is believed to have been to some extent a Christian burial-place. In it a pewter chalice was found which may have been buried with a priest. The objects from these two localities are in the Reading Museum. Numbers of Anglo-Saxon coins have been dug up in Berkshire, more especially in the Cholsey and Wallingford district. They are of silver about the diameter of a sixpence but much thinner and are called pennies.

17. Architecture—(*a*) Ecclesiastical. Churches.

A preliminary word on the various styles of English architecture is necessary before we consider the churches and other important buildings of our county.

Pre-Norman or, as it is usually, though with no great certainty termed, Saxon building in England, was the work of early craftsmen with an imperfect knowledge of stone construction, who commonly used rough rubble walls, no buttresses, small semi-circular or triangular

St Nicholas's Church, Abingdon

arches, and square towers with what is termed "long-and-short work" at the quoins or corners. It survives almost solely in portions of small churches.

The Norman conquest started a widespread building of massive churches and castles in the continental style called Romanesque, which in England has got the name of "Norman." They had walls of great thickness, semi-circular vaults, round-headed doors and windows, and lofty square towers.

From 1150 to 1200 the building became lighter, the arches pointed, and there was perfected the science of vaulting, by which the weight is brought upon piers and buttresses. This method of building, the "Gothic," originated from the endeavour to cover the widest and loftiest areas with the greatest economy of stone. The first English Gothic, called "Early English," from about 1180 to 1250, is characterised by slender piers (commonly of marble), lofty pointed vaults, and long, narrow, lancet-headed windows. After 1250 the windows became broader, divided up, and ornamented by patterns of tracery, while in the vault the ribs were multiplied. The greatest elegance of English Gothic was reached from 1260 to 1290, at which date English sculpture was at its highest, and art in painting, coloured glass making, and general craftsmanship at its zenith.

After 1300 the structure of stone buildings began to be overlaid with ornament, the window tracery and vault ribs were of intricate patterns, the pinnacles and spires loaded with crocket and ornament. This later style is known as "Decorated," and came to an end with the Black Death, which stopped all building for a time.

With the changed conditions of life the type of building changed. With curious uniformity and quick-ness the style called "Perpendicular"—which is unknown abroad—developed after 1360 in all parts of England and lasted with scarcely any change up to 1520. As its name implies, it is characterised by the perpendicular arrangement of the tracery and panels on walls and in windows, and it is also distinguished by the flattened arches and the square arrangement of the mouldings over them, by the elaborate vault-traceries (especially fan-vaulting), and by the use of flat roofs and towers without spires.

Abbey Gateway, Abingdon

The mediaeval styles in England ended with the dissolution of the monasteries (1530–1540), for the Reformation checked the building of churches. There succeeded the building of manor-houses, in which the style called "Tudor" arose—distinguished by flat-headed windows, level ceilings, and panelled rooms. The ornaments of classic style were introduced under the influences of Renaissance sculpture and distinguish the " Jacobean " style, so called after James I.   About this time the professional architect arose.   Hitherto, building had been entirely in the hands of the builder and the craftsman.

Much of the stone used in Berkshire is of local origin, as has already been mentioned in Chapter 12, but a great deal has also been brought from a distance.   Thus it is recorded that when the Abbot of Abingdon in 1100 rebuilt the conventual buildings as well as much of the abbey church, the materials were brought from Wales, six waggons, each drawn by twelve oxen, being engaged in the work.   A great deal of Bath stone will be found in Berkshire buildings and some has even been brought from Caen in Normandy.   Pillars and tombstones of Purbeck marble are common in the churches.   In the south wall of the Dean's Cloisters at Windsor (temp. Henry III) there are clusters of columns and one column in each is of Purbeck marble.

The tower of the church at Wickham, north-west of Newbury, is of a very early style of architecture, showing a variety of " long and short " work.   Two of the belfry windows are double with a pillar in the middle, and are characteristic of this early work.   The walls are very thick.   The remainder of the church has been rebuilt.

On pages 96 and 97 there are views of Norman doorways at Faringdon, both round-headed and one with an embattled moulding over the door.   The church at Avington on the banks of the river Kennet a little below Hungerford is a good example of the Norman style of architecture, and there is a most interesting little church at Finchampstead near Wokingham of which a view is given on page 98. It was built in the twelfth century and the east end of the chancel is round, as was usual at that time. The original windows were probably very small,

North Door, Faringdon Church

and those which we now see were cut in the wall since Norman times. The north aisle, too, is newer than the body of the church, and the brick tower only dates from the seventeenth century. In the church there is a Norman font. There is more or less Norman work remaining in many of our other churches. Thus the illustration on page 92 gives a view of the church of St Nicholas at Abingdon, and a round-headed Norman doorway will be seen under the tower, whilst the remainder of the building belongs to a later style of architecture, probably of the fifteenth century.   The tower of West Shefford

South Door, Faringdon Church

church is curious, the lower part is round and of Norman date, whilst the upper part is octagonal and was built subsequently.

Passing now to the Early English style of architecture there is on page 99 a view of Faringdon Church, which it will be seen is built in the form of a cross with a massive square tower in the middle. Some of the arches inside the church are round-headed like Norman arches, but the windows are of the long narrow shape usual in the Early English style of building. We have churches built mainly

Finchampstead Church

in this style in many places, such as Ardington, Buckland, and Uffington.

Of the Decorated style there is a most beautiful church at Shottesbrook near White Waltham, which was built by Sir William Tressel in 1337. It is cruciform with a tall spire. The walls are of small dressed flints, with corners and window and door frames of stone. The roof is tiled and the spire of stone, the east end window large with beautiful stone tracery (p. 158), and the church is an unusually good example of the Decorated style. The Greyfriars Church at Reading was also built in the Decorated style.   It was long a ruin or used for various purposes, but is now restored.   We also have churches

Faringdon Parish Church

mainly in this style of architecture at Sparsholt, Warfield, and at other places.

We have many examples of Perpendicular style in Berkshire, but by far the best is the Chapel of St George in Windsor Castle (pp. 69, 71).   The greater part of this chapel was built in the time of Edward IV.   The windows are large and the nave consequently very light.   The stone

7—2

roof of the nave was added by Henry VII, and that of the choir by Henry VIII. In the choir are the stalls of the Knights of the Garter, and installation ceremonies of the Order are performed here. St Helen's Church, Abingdon, is our best Berkshire parish church in the Perpendicular style (the tower is Early English). It is large, with five aisles, as will be seen in the illustration

The Upper Cross: East Hagbourne Village

here given. The church at Bray is chiefly celebrated on account of a vicar, one Simon Aleyn, who died in 1588 after holding the living under Henry VIII, Edward VI, Mary, and Elizabeth and altering his views as occasion required. The church is however of itself interesting, and in it will be found examples of Early English, Decorated, and Perpendicular work. The tower belongs to the latest of these styles and is but badly joined on to the aisle of Edwardian date.   It is mostly built of flints, but a broad band of chalk will be noticed about half-way up (p. 149).   There is a good example of a church in this style at Newbury.

Brick church towers are a feature of eastern Berkshire and many of them date from the seventeenth century.

Abingdon Parish Church

One of these, at Finchampstead, is shown in the illustration on page 98.

There are crosses or their remains in many of the churchyards and villages.   At Ardington there is both a new cross and the shaft of an old one.   There are crosses at Denchworth, Goosey, East Hagbourne, Inglesham, North or Ferry Hinksey, Steventon, etc.

At Harwell the rood-screen still remains; there are interesting lead fonts at Childrey and at Long Wittenham; and stands for hour-glasses still exist in the churches at Binfield, Hurst, and Inglesham.

In former times it was very common to keep books in the churches fastened to the shelf or reading-desk by chains, and a few of them still remain. There are several in St Helen's Church, Abingdon. A chained Bible of 1611 is in Cumnor church, and until recently there were several at Denchworth, but they have been removed to the vicarage, and Caxton's *Golden Legend* of 1483 which used to be chained in Denchworth Church is now in the Bodleian Library.

18.   Architecture—(*b*) Religious Houses.

In the year A.D. 528 Benedict of Nursia drew up his celebrated rules at Monte Cassino in Italy, and founded the order of the Benedictine or Black Monks. The order rapidly spread over Europe and was established in Berkshire at an early period. The great Abbey of Abingdon dates from the days of the Saxon Kings, and at the time of Domesday survey it possessed 30 manors in Berkshire besides lands in other counties, and it continued to grow in wealth and power until its dissolution by Henry VIII. The great church of the abbey has been destroyed, but there are some interesting remains of the abbey buildings which, after having been put to varied uses, are now in the hands of the Corporation and carefully preserved.

The illustration given on page 74 shows the south side of what appears to have been a dormitory divided by partitions.

In 1121 Henry I founded a second great Benedictine abbey in Berkshire at Reading, probably upon the site of an older monastic dwelling. Cluny had been founded

Ruins of Reading Abbey

in 910 as an order with a reformed Benedictine rule, and Reading was founded as an abbey of that order. Its connection with Cluny did not, however, last long, and early in the thirteenth century the abbey seems to have become attached to the general Benedictine order. Reading became one of the greatest of English abbeys. Its abbot, like the Abbot of Abingdon, was entitled to wear the mitre and was summoned with the other spiritual peers to attend parliament.

Both Reading and Abingdon were dissolved by Henry VIII, and on November 14th, 1539, Hugh Faringdon, the 31st abbot of Reading, was hanged, drawn, and quartered within sight of his own gateway. The

Part of the Hospitium of St John, Reading Abbey

last abbot of Abingdon had made himself more agreeable to the king, and was granted the manor of Cumnor for life, and a pension as well.

The stone from Reading Abbey was much used for buildings in Reading and the neighbourhood, and in 1556, during the reign of Philip and Mary, a great deal was removed from the abbey and taken by river to Windsor for building the Poor Knights' Lodgings. The inner gateway of the abbey is still standing but has been partially rebuilt in modern times. There are also some remains of the abbey buildings probably belonging to the Hospice of St John.

In the time of William the Conqueror (about 1086) Geoffrey de Mandeville gave the church of St Mary

The Refectory, Hurley Priory

at Hurley, together with certain lands, for a cell of Benedictine monks to be subject to the Abbey of Westminster, and the remains of the priory thus founded are exceedingly beautiful and of much interest. The chapel, built in the Norman style of architecture, is now the parish church of Hurley. The illustration above shows the refectory or dining hall of the priory. The lower part is in the Norman style and the upper part of Edwardian date. On the opposite side of this building is the river Thames.

There was a priory of the Benedictines at Wallingford, and a Benedictine nunnery at Bromhall in the parish of Sunninghill, but there are now no remains of either.

The Abbey Barn, Great Coxwell

The only establishment connected with the great order of the Cistercians in the county was a small cell at Faringdon and a grange or barn at Great Coxwell, both belonging to the Abbey of Beaulieu in Hampshire. The fine abbey barn, dating from the fourteenth century, still remains.

The Austin Canons, an order founded at Avignon about 1061, had priories at Bisham, Poughley, and Sandleford. After the dissolution of the monasteries Bisham Abbey became the seat of the Hoby family. It is beautifully situated on the Thames. Poughley Priory was situated in the chalk district one and a half miles south of Chaddleworth, and there are remains of the buildings at a farm. Sandleford Priory is about the same distance south of Newbury, and some remains are

Bisham Abbey

incorporated in the modern house which was built after plans by Wyatt in 1781 for Elizabeth Montague (1720–1800) the leader of the Blue-stockings.

We have already mentioned Bisham as an abbey of the Austin Canons, founded in 1338, but it had previously been a preceptory of the Knights Templars. That great military order was however suppressed in the time of Edward II and the preceptory dissolved (cir. 1312).

The Templars also had a preceptory at Brimpton which passed into the possession of the other great military order of monks, the Knights Hospitallers. Their chapel, which stands close to Brimpton Manor, still remains and is an interesting building. Shalford farm, a little to the east, was also the property of the Hospitallers. The order was suppressed in England in 1540, and was only temporarily revived under Queen Mary.

There were priories in the county belonging to foreign abbeys and hence termed Alien Priories—one at Steventon belonging to the Abbey of Bec in Normandy, and the other known as Stratfieldsaye, but in Berkshire, belonging to the Abbey of Vallemont, also in Normandy. Both were abolished in the time of Edward III and there are no remains of buildings. A farm named the Priory near Beech Hill occupies the place of the latter, which was on the site of an old hermitage.

There were colleges at Shottesbrook, Windsor, and Wallingford. They were houses of priests who performed divine service in the churches attached to the colleges. We have already mentioned Shottesbrook. There is a very curious alabaster monument to William Throckmorton, one of the later Wardens of the college, in the chancel of the church representing him lying in his coffin.

Besides these religious houses there were houses of Friars at Reading and Donnington, and a number of Hospitals in the county.

19.   Architecture—(c) **Military.**

Attention has already been drawn to the earliest fortifications in the county.   They were banks of earth and
had probably wooden palisades.   In Norman times fortified
residences became common and were usually of stone.
The history of a Norman castle was probably often as
follows.   In the first place a tower called a "keep" was
built and was protected by a moat and probably by some
earthworks.   Then at a later date the earthworks were
replaced by walls, which usually enclosed a larger space
than the older fortification.   The walls were usually
strengthened by towers, but the keep still remained the
citadel of the fortress.

We know that William the Conqueror built a castle
on the chalk hill at Windsor before the year 1086, but
we know nothing of its plan or form, for no part of the
present castle can be dated before the reign of Henry II,
and even of that time there are only the foundations and
part of the lower story on the south side of the Upper
Ward.   The imposing western wall of the Lower Ward,
with its three towers, belongs to the time of Henry III.
The Round Tower on its high mound is the keep of the
castle, and much of it is as old as the time of Henry III.
The top part, however, is modern.   Close to the Round
Tower is an old Norman gate which was rebuilt by
Henry III and again by Edward III.   The gateway could
be closed by doors and also by a portcullis or grille let
down from above, and the portcullis is still in its place
ready to be lowered.

The view of Windsor Castle given on p. 2 is taken
from the Buckinghamshire bank of the river Thames
and shows the north side of the castle.　On the left are
the buildings which contain the state apartments; in the
centre is the Round Tower.　To the right we see

The Round Tower, Windsor Castle

St George's Chapel with its great west end window, and
still further to the right is the Clewer or Curfew Tower
with the pointed roof.　The main part of this tower
dates from Henry III, and it has been used as a bell tower
since the time of Edward III.　The pointed roof is

modern. St George's Hall (p. 78) is in the part of Windsor Castle known as the State Apartments, and in it the feasts of the Knights of the Garter are held. It is an old hall, but was much altered by Sir Jeffry Wyatville, the architect employed by George IV to repair the castle. There was nearly always a well in the keep of a Norman castle, and this was the case at Windsor, the well in the Round Tower being 160 feet deep.

There is a great rectangular earthwork at Wallingford which may go back to Roman or early British times, but in any case it was adopted by the Normans and a castle was built on the site. The mound on which the keep was built still exists, but little else of these buildings survives.

No remains of the castle at Newbury exist. It stood on the south bank of the river Kennet and was built about 1140. The mound upon which the keep stood is all that we have left of the castle of the St Walerys at Hinton Waldrist, and a moated enclosure by the side of the river Loddon is all that remains to mark the site of the castle named Beaumyss, built by one of the De la Beche family in 1338.

Of Donnington Castle near Newbury we have the remains of some walls and a gateway with two round towers. The walls are mainly flint with some stones of various sorts intermingled. There are stone courses and stone door and window frames. Repairs have been made with brick. The castle was built in the time of Richard II. It stands upon a hill or spur which runs out in a southerly direction from the plateau named

Snelsmore Common, and it overlooks the valley of
Newbury. On the west and south there is a steep slope
down towards the river Lambourn, and on the east is
a deep valley in the chalk. On the north the slope up
to the Common is gradual, and so the position is a very

Gateway, Donnington Castle, Newbury

strong one. Donnington Castle played an important
part in the Civil War of 1642–9, and underwent a long
siege in 1644–6.

In former times dwelling places, even though not
fortified, were at least protected by a moat. The inter-

esting old manor house of Ashbury is still moated on
three sides, and the old moat remains in a more or less
perfect state round many a farm in the county.

20. Architecture—(*d*) Domestic.

The churches of the eleventh and succeeding centuries
which remain are well adapted for their use now, but
this cannot be said of the dwelling-houses of Norman or
Edwardian landowners, and this is one reason why we
have but few left in anything like perfect condition.
The residence of the chief landowners of the twelfth
century, when not a castle, consisted of a hall, usually on
the ground floor, but sometimes with a lower story half
below the surface level, and the hall was not only a
reception and dining room, but was also the sleeping place
for the greater number of the persons living in the house.
In many cases there were, no doubt, subsidiary chambers,
which might serve as more or less private apartments for
the landowner himself, and as time went on the number
of the subsidiary chambers increased and the importance
of the hall diminished, but it impressed itself so firmly on
the popular mind that the word still remains in use for
the house of the landowner, which is often spoken of as
"the Hall."

There is a doorway belonging to a hall of the Norman
period at Appleton in the northern part of the county,
and we have already noticed some remains of the residential buildings of the monks of Abingdon, belonging to the

thirteenth century. At Charney, about seven miles to the west of Abingdon, there are some interesting remains of a building which was occasionally the residence of the Abbots. The private chapel and much of the house are still standing. These buildings, known as the " Monks House," date from the thirteenth century and are incorporated in a modern house.

There are two old houses at Sutton Courtney south of Abingdon. The one is opposite the tower of the church, and is of Norman and Early English style, the second is a manor house of the time of Edward III, the hall of which, with its roof and windows, has been very little altered. Cumnor Hall has vanished, excepting a fragment of wall, but some of the windows and a doorway are still to be seen in Wytham Church.

It has been mentioned that one reason why few old dwelling-houses remain is that they would not be suited to modern requirements, but another reason is that they were often built of wood. In the fifteenth century buildings of timber and brick became common, and some of them remain at the present day. Ockwells, rather more than a mile south-west of Maidenhead station, was probably built in the time of Edward IV. It was for some time the residence of the Norris family (see page 138). The house was not fortified, and is of timber and brick with a tiled roof. One may gain a good idea of the appearance of the dwellings of our ancestors in Tudor times from the Horseshoe Cloisters in Windsor Castle, though they were practically rebuilt recently by Sir Gilbert Scott. Timber and brick farmhouses and cottages may be seen all over the county, belonging to all dates from the Tudor times to the present day.

Many of the most beautiful private houses in England were built during the reign of Queen Elizabeth, and we

Cottage at Cookham Dean

have some examples in Berkshire.   Shaw House, about a mile north-east of Newbury, was built in 1581.   It is of red brick, with tall brick chimneys and a tiled roof.   The corners of the house and the window and door frames are of stone, and in fact there is a good deal of stone.   The house was occupied by Charles I on the day of the second

8—2

battle of Newbury, October 27th, 1644, and the remains
of earthworks thrown up by his troops are still to be seen
in the garden. Billingbear, near Binfield, is an Elizabethan
house standing in a large and beautiful park.

Ufton Court, near Aldermaston, was built in the latter
part of the sixteenth century. Farmhouses of the same

Wayside Cottages, Bisham

period are to be seen at Lyford, west of Abingdon, East
Hendred, Great Coxwell and at other places.

Secret rooms are often to be found in old houses.
There is an example at Bisham Abbey, with a fireplace,
the chimney of which is said to be connected with that of
the hall, so as to prevent its smoke being observed. At
Ufton Court there are several hiding-places, one of which
has an exit to the open air. It is said that Charles I

passed the night of November 19th, 1644, in a secret room at the manor house, West Shefford.

In 1852 some houses which stood on the site of the former ditch of Windsor Castle were removed, and a passage was found cut through the chalk, with stone steps and stone arching. It had probably been a secret way from the interior of the Castle to the moat.

We have many buildings in Berkshire belonging to the seventeenth century. Coleshill House, south-west of Faringdon, was built by the celebrated architect Inigo Jones (1572–1652) at the time of the Commonwealth, and he also built most of Milton House, near Steventon, in which village are some beautiful old houses. Buscot House, in the north-west corner of Berkshire, is an example of the comfortable, though not very beautiful mansions built at the close of the eighteenth century. The residential part of Windsor Castle dates in part from the reign of Henry II, but it has been greatly altered from time to time. Its present appearance is largely due to Sir Jeffry Wyatville (1766–1840), who modified and rebuilt a great deal in the time of George IV. His object was to make the Castle a comfortable residence and at the same time to preserve the appearance of an ancient fortress.

21. Communications — Ancient and Modern.

The Ridge Way is one of the oldest roads in England. It enters Berkshire on the chalk downs above Ashbury at a level of 600 feet above the sea, and runs in an easterly direction by Wayland Smith's Cave and Uffington Castle; thence by Hackpen Hill to Letcombe Castle, along the top of the ridge north of West and East Ilsley. From

The London Road near Sunninghill

here, turning to the right across the little valley on Compton Downs, the road probably reached the river Thames at Streatley. This old road is also known as the Icknield Way, and there is another old road named the Port Way, which follows the valley north of the chalk downs, running through Ashbury and Wantage. It is marked on the maps as a Roman road, and probably both roads were in use in Roman times, though the Ridge Way at least is almost certainly of much older date.

The Roman road from Marlborough to Silchester followed much the same line as the modern road from Hungerford to Speen near Newbury, but there does not seem to be any trace of the road from that place to Silchester. The Roman road from Cirencester to Silchester ran by way of Baydon and Wickham, joining the Marlborough road at Speen. There is but little trace of the Roman road from Silchester to Dorchester in Oxfordshire, but the Silchester and London road is fairly well marked, and part of it, as we said in a former chapter, is known as the " Devil's Highway."

In the middle ages the roads were exceedingly bad, and even in the seventeenth century they were far from satisfactory. Pepys mentions, in his *Diary*, June 16th, 1668, that he lost his way driving from Newbury to Reading. This, it will be observed, was in the summer, and one would think on a well-known road.

In the eighteenth century the roads were gradually improved, and towards the end of the century began to be kept in good order for the coaches, which were also rapidly improving.

In the early part of the nineteenth century two mail coach routes ran through Berkshire.

The road from London to Gloucester entered Berkshire at Maidenhead and left the county at Henley.

After passing through Oxford it again entered Berkshire, and ran by Cumnor and Fyfield to Faringdon. From that place it ran by Buscot Park and crossed the river Isis at St John's Bridge near Lechlade.

The London and Bath mail route ran through the county by Maidenhead, Reading, Newbury, and Hungerford.

Hungerford Canal

Besides these mail-coach routes there were several roads in the county which came under the head of "turnpike roads." The term turnpike road means a road having toll-gates or bars on it. The toll-gates were first constructed about the middle of the eighteenth century, and were called turns, and the turnpike road was one upon which those who refused to pay toll could be turned back. Turnpike roads are now practically extinct and a new

Hambleden Weir

species of highway called main roads has taken their place. The cost of repair is borne partly by the county and partly by the Local Highway Authority.

Canals and Rivers. Canals have to a large extent been superseded by railways in these days. It is, however, possible that the advent of cheap motor traction may cause

Disused Canal between Abingdon and Wantage

them to revive. The Kennet and Avon Canal runs from Newbury, and entering Wiltshire near Hungerford furnishes a waterway from the Thames to the Severn. The navigation of the river Thames is improved by a number of weirs and locks, most of which have been re-made in recent times, and if more useful they are much less picturesque than in former days. The level of the

river at Hambleden weir is just about 100 feet above the
sea. The river Kennet is also provided with a series of
weirs and locks. A canal which ran from Wantage to
Abingdon is now disused.

The bridges over the rivers are for the most part
modern, but many of them replace older structures,
indeed most of the crossing-places are very old. The

Boulter's Lock

bridge at Abingdon was originally built in the fifteenth
century, and was under the charge of the Guild of the
Holy Cross, and Maidenhead Bridge was the property of
a corporation from early days.

Railways. The Great Western Railway enters Berkshire at Maidenhead, and runs by way of Reading to near
Goring, where it crosses the Thames into Oxfordshire,

returning into Berkshire near Moulsford. It then passes
by way of Didcot into Wiltshire, which county is entered
a little before the line reaches Swindon. An important
branch of the Great Western runs from Didcot to Oxford,
and another branch of the same railway from Reading to
Newbury, Hungerford, etc. Express trains to the west
of England pass over both the Didcot and the Newbury
line, and in these days they are frequently run from
Paddington to far beyond the Berkshire border without
a stop.

The Great Western has branch lines to Windsor, to
Cookham for High Wycombe, to Henley, to Wallingford,
to Abingdon, and to Faringdon, and also a rather important line from Reading to Basingstoke, giving a
communication from Oxford to the south coast. There
is also a light railway with auto-cars running between
Newbury and Lambourn which belongs to the Great
Western.

The Didcot, Newbury, and Southampton Railway
runs from the first-named place in a southerly direction,
crossing the Reading and Newbury line at right angles.

The South Eastern and Chatham Railway Company
have a branch line running to Reading. It enters Berkshire near the village of Sandhurst.

The London and South Western Railway have
branches to Windsor and to Wokingham, and from the
latter place run trains over the South Eastern line to
Reading.

22.   Administration and Divisions— Ancient and Modern.

The division of the county into Hundreds dates from Saxon times.   Each Hundred was governed by a High Constable, or Bailiff, and formerly there was a Court of Justice, called the Hundred Court, which was held regularly for the trial of causes, but this court fell into disuse.   By various Acts of Parliament the Hundred is made liable for damage caused to persons by riots.

In early days most of Berkshire was divided amongst different manors, and each manor had a Manorial Court or Court Baron.

It has been already explained in Chapter 3 that the present administrative county differs somewhat from the geographical county, and as the town of Reading with a tract around it has been formed into the "County Borough of Reading" it is not for most administrative purposes a part of the county of Berks.

The chief officials of Berkshire, under His Majesty the King, are the Lord Lieutenant, the Custos Rotulorum, and the Sheriff.   The first two of these offices are usually held by the same person.

The office of Lord Lieutenant dates from about the time of Edward IV, and he was formerly the chief military officer of the Crown in the county.   The Custos Rotulorum is the first amongst the justices, but the High Sheriff has precedence in the county.   The Custos selects the county magistrates, and they are appointed by the

Lord Chancellor. The office of Custos dates from the time of Edward III. He is nominally the keeper of the County Records, but in these days they are in fact in the charge of the Clerk of the Peace.

The Sheriff was originally elected by the people in the county, but since the time of Edward II he has been appointed by the Crown. He was the agent through whom the King collected his dues, and in time became the military as well as the judicial and executive head of the county and headed the *posse comitatus* or power of the county. During the Wars of the Roses his influence became less, and the Lord Lieutenant took his place to some extent. It is recorded in Domesday Book that Godric the Sheriff of Berkshire gave a lady with the name of Aluuid half an acre of the royal domain as a present for teaching his daughter the art of gold embroidery.

The Sheriff is the first man in the county, taking precedence of all peers and of the Lord Lieutenant. He is appointed annually.

The county is divided into Petty Sessional divisions for magisterial purposes, and the Court of Quarter Sessions is a general meeting of all the justices of the county. In boroughs, Reading, Abingdon, Newbury, and Windsor, the Court of Quarter Sessions is held by a Recorder.

The affairs of the county (not including Reading, which is a County Borough of itself) are managed by the County Council, which was established by statute of 1888, and by District and Parish Councils, which were established by an Act of 1894. The county is divided into eleven districts, Bradfield, Windsor, Cookham,

Easthampstead, Wokingham, Newbury, Hungerford, Wantage, Wallingford, Faringdon, and Abingdon.

The affairs of the County Borough of Reading are managed by its Mayor and Corporation.

The Town Hall, Abingdon

For purposes of Assizes, Berkshire is on the Oxford circuit, and the Court is held at Reading.   County Courts are held from time to time at the various towns.   The

County Court circuits are quite different from the Assize Court circuits.

For Parliamentary elections the county is divided into three divisions, Abingdon, Newbury, and Wokingham, each of which returns one member to Parliament. Reading also returns a member to Parliament, and so does Windsor, but the Parliamentary borough of Windsor includes a considerable tract outside Berkshire.

23. Public and Educational Establishments.

The municipal buildings at Reading were erected during the period 1875–1897, and consist of two Town Halls, the Borough Council offices, a Free Library, the Museum, and an Art Gallery. On the walls of the reading-room there is a good collection of views of Reading and of the river Thames.

The Town Hall at Windsor was built by Sir Christopher Wren. On the exterior there are statues of Queen Anne and her husband, Prince George of Denmark. The Town Halls at Wokingham and Newbury are modern brick buildings. The Cloth Hall at the latter place, now a museum, is very interesting. It was built by the Guild of Clothworkers of Newbury, which was incorporated in 1601, and has a picturesque wooden cornice and wooden pillars, and a red tiled roof.

The Town Hall at Wallingford dates from 1670, and is supported by pillars, leaving an open undercroft. The

Abingdon Town Hall has also an undercroft and dates from 1677. It is said, however, to have been designed by Inigo Jones, who died in 1652. There is an interesting old Town Hall at Faringdon.

The Cloth Hall, Newbury

The Royal Berkshire Hospital at Reading was opened in 1839, and there are many hospitals, homes, and orphanages in various parts of the county.

The Prison at Reading stands upon part of the site of Reading Abbey. There is a large County and Borough lunatic asylum at Moulsford, and a very large criminal lunatic asylum at Broadmoor, in the eastern end of the county.

The Royal Military College, Sandhurst, is one of the chief Government institutions for the education of officers

The Town Hall, Wallingford

for the army. It was built in 1812, and though quite plain in style, the long frontage on a rising ground, above a fine lake, is distinctly effective. The Staff College is in the same grounds, but is in Surrey. Considerable additions are now (1910) being made to the buildings at the Military College.

Royal Military College, Sandhurst

Wellington College, also near Sandhurst, was built as a public school by public subscription in memory of the great Duke of Wellington, who died in 1852. By the end of 1858 a sum of £145,785 had been received. This included a grant of £25,000 from the Patriotic Fund. The buildings are of red brick with stone corners, etc., and were completed in 1859. They have, however,

The Town Hall, Faringdon

been greatly added to since. The chapel is by Sir Gilbert Scott. The first head master was Edward White Benson, who subsequently became Archbishop of Canterbury.

Bradfield College is another important public school, founded by Thomas Stevens in 1850. The buildings are of red brick and flint, and are partly old. There is an open-air theatre where Greek plays are performed.

Radley College is beautifully situated by the river Thames. The site was part of the property of the Abbots of Abingdon, and passed through the hands of the families of Stonehouse and Bowyer. Much of the old

Gate of the Old Grammar School, Abingdon

mansion is incorporated in the college buildings. The college was founded by the Rev. William Sewell, D.D.

University College, Reading, is a comparatively new establishment, and the buildings are still in process of construction. Higher teaching in literary and scientific subjects is given, and there is an Agricultural Department, a Dairy Institute, and a Horticultural Department. There has been a school at Reading from quite early times, but its history has been a somewhat broken one. In 1783 John Lempriere published his *Classical Dictionary* whilst an assistant master at the school, and Richard Valpy was its head master for 55 years (1781–1836).

In addition to the above there are several important recognised secondary schools at Abingdon, Bracknell, Clewer, Maidenhead, Newbury, Wallingford, Wantage and Windsor.

There are many almshouses in Berkshire, the most interesting of which is Christ's Hospital, Abingdon. It is of brick and timber with an open gallery (p. 63). It was founded under its present name by Charter of Edward VI, but had a previous existence. The almshouses near Wokingham, built 1663, and known as Lucas Hospital, are a good example of seventeenth century brickwork, and are very picturesque. The Jesus Hospital at Bray was founded in 1627 for 40 poor persons. It is a most attractive red brick building, with a quadrangle in the middle, and a small chapel, the windows of which have stone frames which were probably taken from an older building. The quadrangle is shown in the picture by Frederick Walker in the Tate Gallery, named "The Harbour of Refuge."

24. The Forest in Berkshire.

Windsor Forest consisted in early times of a tract of wood and heath which even before the Norman Conquest was looked upon as Crown property. It is one of the five forests mentioned by name in Domesday. It was no doubt of great extent, but its boundaries are not known even if they were ever very clearly defined. There is in the British Museum a volume of maps and plans of Windsor by John Norden, made in the early part of the reign of James I, and the map of the forest shows that it was at that time bounded by the Thames on the north, by the Loddon on the west, by the Blackwater on the south, and that it extended to the east into Surrey as far as the Hog's Back, Guildford, and the river Wey.

The forest was at that time divided into 16 walks, each under a Keeper. Two of these, Cranbourne and New Lodge Walks, appear to have been previously known as Cranbourne Chase, and together with Egham Walk were the part of the forest lying nearest to the castle, and including what is now the Home Park and the Great Park. The other walks in Berkshire were Swinley Walk, Easthampstead Walk, Sandhurst Walk, Bigshot Walk, Bearwood Walk, and Warfield Walk. There was also a large district extending from Maidenhead and Bray to Wokingham and Twyford, which was called the Fines Bayliwick, and of which Sir Henry Neville claimed to be Keeper by inheritance.

Several parks are marked in the forest. Of these the

Little Park is now the Home Park, Windsor, and the Great Park and Moat Park are in the present Windsor Great Park. Sunninghill Park, Foliejon Park, Easthampstead Park, and Bagshot Park, the last mostly in Surrey, still remain.

Besides the Parks there were certain enclosed places called Rails. Cranbourne Rails is in Windsor Park.

Ascot Race Course

Swinley Rails was until recently the place where the deer for the Royal Hunt were kept, and Bigshot Rails is apparently the place now named Ravenswood, near Wellington College.

In the early part of the nineteenth century there was a great deal of discussion as to the rights of the Crown over Windsor Forest, and in 1813 an Act of Parliament was passed dealing with the matter, and the Forest is now enclosed either as Crown land or as the property of private persons. Ascot Race Course is in the old Swinley Walk.

Walter Fitz Other was appointed by William the Conqueror Castellan, or Governor of Windsor Castle, and Warden of the Forest; and the office, which has become known as that of Constable of the Castle, has existed from his appointment to the present day.

25. Roll of Honour.

King Alfred was born at Wantage in the year 849, and his statue by Count Gleichen stands in the market place. The exact site of the palace of the Kings of Wessex, in which he was born, is not known, probably it was a wooden building. Edward III and Henry VI were both born at Windsor; Henry I was buried at Reading; Henry VI, Edward IV, Henry VIII, Charles I, George III, George IV, and William IV were buried at Windsor; and Queen Victoria and the Prince Consort lie in the mausoleum at Frogmore, in Windsor Park. King Edward VII was buried at Windsor May 20, 1910.

The Marshals of Hampstead Marshall were a family of warriors. The most distinguished of them was William, first Earl of Pembroke. When he was a child his father, John Marshal, was besieged at Newbury by King Stephen, 1152, and William was given as a hostage for a truce and the surrender of Newbury Castle. The father did not keep his terms, and the child would have been killed had not Stephen taken a liking to him and saved his life. He

became a great soldier and served Henry II, Richard I,
John, and Henry III with the utmost fidelity, becoming
Regent of England during the early part of the reign of
Henry III. He died in 1219 at Caversham, and is buried
in the Temple Church in London.

In later times another warrior owned Hampstead
Marshall. This was William Craven, Earl of Craven
(1606–1697). He fought in the German wars of 1632–37
and was the faithful champion of Elizabeth Queen of
Bohemia, the only daughter of James I. At the
Revolution of 1688, though over 80 years old, he was in
command of the King's Guards, and Macaulay, in his
History of England, describes how unwillingly the stout
old soldier made way for the Dutch troops at Whitehall.
Ashdown Park was another seat of the Earl, and is still
in the possession of his descendant.

Radley belonged to a gallant sailor, Admiral Sir
George Bowyer, Bart. (1740?–1800), who lost a leg off
Ushant, June 1st, 1794. Another Admiral, Samuel
Barrington (1729–1800), is buried at Shrivenham. He
served under Hawke and Rodney, and was commander-
in-chief in the West Indies.

The family of Norris or Norreys has long been connected with Berkshire. Richard de Norreys, a member
of a Lancashire family, held the office of cook to Eleanor,
wife of Henry III, and in 1267 the manor of Ockholt,
near Maidenhead, was granted to him. One of his
descendants, John Norris, who held office in the Court
of both Henry VI and Edward IV, built the house
Ockwells at Ockholt, which has been already mentioned

on page 114. He was buried at Bray in 1467. One branch of the family settled at Fyfield, and another branch became Norris of Rycote, which is in Buckinghamshire, but they too held much Berkshire property. Hampstead Norris derives its second name from this family. Henry

Archbishop Laud

Norris was an intimate friend of Henry VIII, and was present at the Field of the Cloth of Gold. He, however, fell under the suspicion of being a lover of Anne Boleyn, and was in consequence executed in 1536. His son, also named Henry, was created Baron Norris of Rycote in

1572. He died at Englefield, and there is a monument to him and to one of his six soldier sons, Sir John Norris, in Yattendon church. Francis Norris, a grandson of Henry Lord Norris, was born at Wytham, and in 1621 was created Earl of Berkshire. He left no sons, and the earldom became extinct at his death, 1623. The barony descended through two ladies, Elizabeth and Bridget, to James Bertie, who was created Earl of Abingdon 1682. The present peer, whose seat is Wytham Abbey, is the seventh Earl of Abingdon.

St Edmund (1170?–1240), Archbishop of Canterbury, was born at Abingdon, and William Laud (1573–1645), also Archbishop of Canterbury, was born at Reading, the only son of William Laud, a clothier. He was educated at the Free School at Reading, and he gave a farm to Reading for charitable purposes. It was sold a short time ago, and the purchase money invested, producing some £330 a year. Another charity at Wokingham established by him also still exists.

John Jewel, Bishop of Salisbury (1522–1571), was for some time vicar of Sunningwell. He was a voluminous writer on theological subjects. Another churchman connected with Sunningwell was John Fell (1625–1686), Bishop of Oxford, who was born either there or at Longworth. His father was rector of the parish. In 1648, at the time of the Civil War, he was turned out of his Studentship at Oxford, but continued to celebrate the rites of the church in a house opposite Merton College. He was a distinguished man, but is best known by the lines referring to him which begin "I do not love thee, Doctor Fell." Joseph Butler (1692–1752), Bishop of Durham, and the author of the *Analogy of Religion Natural and Revealed to the Constitution and Course of Nature*, was born at Wantage, the son of a retired draper who lived at the Priory.

Sir Philip Hoby (1505–1558) and his half-brother, Sir Thomas Hoby (1530–1566), were both distinguished

The Hoby Chapel, Bisham Church

diplomatists. The former received the manor of Bisham from Henry VIII, and they are both buried there. Queen Elizabeth was domiciled at Bisham under the charge of the Hobys for a time during the reign of her sister Mary.

Sir John Mason, another diplomatist of the same period, was the son of a cowherd at Abingdon. He is described as a paragon of caution, coldness, and craft, and held high office, diplomatic and political, under Henry VIII, Edward VI, Mary, and Elizabeth, being in favour with all these sovereigns.

Sir Henry Unton, or Umpton, who died in 1596, was both diplomatist and soldier of the time of Elizabeth. He was born at Wadley Hall, near Faringdon, where the Queen visited him in 1574. The house is still standing. There is a fine alabaster monument to him in Faringdon church.

In 1626 the title of Earl of Berkshire was conferred on the Hon. Thomas Howard, of Charlton, Wilts. He was a son of the Earl of Suffolk, and in 1745 the two titles passed to one man, and are so held at the present day.

William Lenthall (1591–1662), the Speaker of the House of Commons in the Long Parliament, bought Besselsleigh, the house of the ancient family of Besils, and his descendants still own it.

William Penn (1644–1718), the Quaker and founder of Pennsylvania, though London-born, lived at Ruscomb, near Twyford, for some time towards the end of his life and died there.

Passing now to authors, Henry Hallam (1777–1859), the historian, was born at Windsor, the son of one of the canons. Catherine Sawbridge (1731–1791), who became in turn Mrs Macaulay and Mrs Graham, was the authoress of a History of England. In her later years she lived at Binfield and is buried in the churchyard there. The antiquary, Thomas Hearne (1678–1735), was the son of the parish clerk at White Waltham, and was born at Littlefield Green.

Jethro Tull (1680–1741), a well-known writer on agriculture, was born at Basildon, and farmed land first near Wallingford, then in Oxfordshire, and finally near Hungerford. About the year 1701 he invented a horse-drill for sowing seed. He is buried at Basildon.

Sir William Blackstone (1723–1780), of the *Commentaries*, was buried in St Peter's, Wallingford, at which place he had spent much of the latter part of his life. John Shute, first Viscount Barrington (1678–1734), author of the *History of the Apostles*, lived at Beckett House, Shrivenham, which was left to him by Sir John Wildman.

Several poets were connected with Berkshire, but chief of them all was Alexander Pope (1688–1744), whose father owned a small property at Binfield. Here the poet lived for much of the early part of his life. His poem *Windsor Forest* contains many lines dealing with the district around Binfield. Sir William Trumbull (1639–1716), the friend of Pope, and Secretary of State in 1695, lived at East-hampstead, not far from Binfield; and Elijah Fenton (1683–1730), another of Pope's friends, himself a poet, lived with the Trumbull family during his last years.

Henry James Pye (1745–1813), though Poet Laureate, wrote but poor verses. He commuted the tierce of canary to which the Poet Laureate was entitled for £27 a year. He was a son of Henry Pye of Faringdon, and at one time was M.P. for Berkshire. Joshua Sylvester (1563–1618), also a poet, is said to have lived at Lambourn as steward to the ancient family of Essex, and one of his volumes is dedicated to Mistress Essex of Lambourn.

Mrs Elizabeth Montague (1720–1800), whose London house was a centre of intellect and fashion, where the term "Blue-stocking" was first applied to her conversation

Miss Mitford

parties, lived a good deal at Sandleford Priory, near Newbury, and built a large house there from plans by Wyatt.

She cannot, however, claim the close connection with Berkshire, both as regards life and writings, which is so characteristic of Mary Russell Mitford (1787–1855) who lived for a time at Reading, then at Three Mile Cross, and finally at Swallowfield, in the churchyard of which place she lies buried. Her best known work is *Our Village*, the scenes in which are laid in the district at and around Three Mile Cross.

Thomas Day (1748–1789), the author of *Sandford and Merton*, was the owner of Bear Hill, Wargrave.

John Winchcombe, alias Smalwoode (died 1520), was a pioneer of the clothing manufacture at Newbury, and acquired thereby great wealth. He built a house at Bucklebury on land which had belonged to the Abbey of Reading. His descendant, Frances Winchcombe, married in 1700 the celebrated Viscount Bolingbroke, who resided at Bucklebury for a time. John Winchcombe is buried in Newbury church. He was popularly known as "Jack of Newbury" and many fables are told about him. Thomas Deloney, a weaver by trade, who lived in the latter part of the sixteenth century, wrote the ballad "The Pleasant History of John Winchcomb, in his younger days called Jack of Newbury."

26. THE CHIEF TOWNS AND VILLAGES OF BERKSHIRE.

(The figures in brackets after each name give the population of the town or parish in 1901, and those at the end of the sections give the references to the text.)

Abingdon (6441). A municipal borough in the Abingdon division of the county. It is situated at the junction of the river

Abingdon Bridge

Ock with the Thames, 61 miles from Paddington by railway, and 56 miles from London by road. It was incorporated by Charter granted by Philip and Mary in 1555. Its trade is mainly in agricultural produce, and its manufactures are carpets, woollen goods, and sacking. We have already referred to the remains of its great Benedictine Abbey as well as to its churches, Christ's Hospital, and the Town Hall. The Earl of Abingdon is the High Steward of the borough. (pp. 19, 24, 34, 57, 62, 65, 74, 92, 94, 97, 100, 101—4, 113, 116, 122, 123, 129, 134, 140.)

Aldermaston (482). A village with a railway station eight miles from Reading on the Newbury line. The church is of various styles. There is a Norman doorway built in at the west end under the tower. The east window of three lights is Early English. (p. 116.)

Aldworth (211). A village on the chalk downs three miles west of Streatley. The church is celebrated for the series of tombs of the De la Beche family with effigies and canopies of the Edwardian period.

Appleton (466). A village near the Thames five miles north-west of Abingdon. The remains of a Norman manor house exist near the church. It is defended by a moat, and there are two other moated houses at no great distance. (pp. 18, 113.)

Ardington (433), a village at the north side of Lockinge Park with a church mainly in the Early English style. There is a fine chancel arch, and the north doorway is round-headed. (pp. 78, 101.)

Ascot Heath (1927). A village and parish with a railway station 29 miles from Waterloo. The race-course is close to the station. (pp. 16, 39, 136.)

Ashbury (589). A village five miles north-west of Lambourn; the church with some windows in the Decorated style, a good Norman doorway, and other points of interest. In the parish there is a manor house of the fifteenth century moated on three sides. The area of the parish is 5609 acres and the population has been reduced from 786 in 1851 to 589 in 1901. (pp. 84, 113, 117, 119.)

Avington (97).   A village on the river Kennet two and a half miles east of Hungerford.   It has a very curious and fine Norman church with a rich arch between the nave and the chancel.   The font with 13 figures is Norman.   (p. 95.)

Balking (295).   A village in Uffington parish, and near Uffington station.   The church is small with a very good Early English chancel, and an east window of three lancet lights.

Basildon.   (pp. 90, 142.)

Beech Hill (265).   (p. 108.)

Binfield Rectory

Beedon (232).   A scattered village or hamlet in the chalk district south-west of Compton.   The church belongs to the period of transition between the Norman and Early English styles.   The font is Early English.

Binfield (1892).   A village and district three miles north-east of Wokingham, the early home of the poet Pope.   The church is largely built of conglomerate from the gravel.   The arch under the tower, Perpendicular in style, is of chalk.   In the church there is a chained copy of the Paraphrase of Erasmus upon the New Testament. Billingbear, a fine Elizabethan house with a large park, lies to the north-west of the village. (pp. 101, 116, 142, 143.)

Bisham (594). A parish on the Thames a little above Cookham. The church and abbey have been already referred to. (pp. 57, 73, 76, 77, 106, 107, 116, 141.)

Bray Church

Boxford (461). A village with a railway station on the Lambourn line four miles north-west of Newbury. Many Roman remains have been found in the parish.

Bradfield (1526). A village seven miles to the west of Reading. Bradfield College is a well-known public school. (pp. 21, 132.)

Bray (1722). A village on the Thames between Maidenhead and Windsor. The well-known vicar, Simon Aleyn (died 1588) succeeded in retaining his living during the reigns of Henry VIII, Edward VI, Mary, and Elizabeth. The song wrongly gives him a later date. The church is partly Early English; the tower is Perpendicular. Bray gives its name to the Hundred, which includes most of Maidenhead. (pp. 20, 62, 100, 134, 138.)

Buckland (665). A large village four miles north-east of Faringdon. The large cruciform church is mostly Early English. The central tower is low and massive with fine Early English tower arches. The tracery has in modern times been removed from most of the windows. The population of the parish has diminished in recent years. (pp. 88, 98.)

Bucklebury (1066). A village in a large parish six miles north-east of Newbury. Swift visited Henry St John, Viscount Bolingbroke, at Bucklebury in 1711. (p. 21.)

Burghfield (1352). A village in the clay district five miles south-west of Reading. A curious wooden effigy of the fourteenth century is preserved in the church. (p. 57.)

Chieveley (1204). A village four miles north of Newbury. The church is partly in the Early English style, the chancel with good lancet windows. The south doorway is round-headed and late Norman. Cromwell is said to have slept at the Old Blue Boar Inn the night before the second battle of Newbury.

Cholsey (1826). A large village with a railway station 48½ miles from Paddington, the junction for Wallingford, distant 2½ miles to the north-east. The large cruciform church has a fine Early English chancel. The arches of the central tower are massive and early Norman, and there are good Norman doors and windows in the church. The upper part of the tower belongs to the Decorated period. (p. 91.)

Clewer (6171) on the river Thames is practically a suburb of Windsor, with numerous orphanages, homes, and other charitable institutions.

Coleshill (342). A village three and a half miles west-south-west of Faringdon on the Berkshire side of the river Cole. Coleshill House was built from designs by Inigo Jones. There are late Norman and also Early English arches in the church and the tower with its parapet and pinnacles is a good example of the Perpendicular style. The base and shaft of a village cross remain in the churchyard. (p. 117.)

Cookham Lock

Cookham (3007). A village with a railway station on the Thames a little above Maidenhead. The church is largely Early English in style, the solid square tower is Perpendicular and is a prominent object from the river. (p. 17.)

Coxwell, Great (264). (pp. 106, 116.)

Crowthorne (3185). A village and ecclesiastical district in the parish of Sandhurst. On Norden's map of Windsor Forest (temp. James I) the name is given to a tree at a point where three of the Walks met, and the place is also on the boundary of three parishes. Wellington College and Broadmoor Lunatic Asylum are close to the village.

Cumnor (870). A village three miles south-west of Oxford. The church is late Norman and Early English with some later work. The tower has a round-headed west doorway and good Transition tower arch. There are scarcely any remains of Cumnor Hall. (pp. 77, 102, 104, 114, 120.)

Didcot (420). An important junction on the Great Western Railway 53 miles from Paddington. In the church is an effigy of the thirteenth century with a mitre, supposed to be that of the first mitred abbot of Abingdon. The base of the cross in the churchyard is old. (pp. 88, 124.)

Donnington. A hamlet two miles north of Newbury, with a castle and priory. (pp. 80, 108, 111.)

Earley (10,485), is becoming a suburb of Reading. Whiteknights, a seat of the 4th Duke of Marlborough, has now vanished and the park is partly built over.

Easthampstead (1708), a village three and a half miles south-east of Wokingham, gave its name to one of the Walks in Windsor Forest. Caesar's Camp (see page 89) is a mile to the south. There are four windows by Burne Jones in the church. (pp. 89, 90, 135, 143.)

Englefield (315). A village and park five miles west of Reading. (pp. 77, 139.)

Faringdon (2770). A market town with railway station 70 miles from Paddington. The trade is mainly in cattle, sheep, bacon, and corn. (pp. 35, 62, 67, 70, 88, 95, 97, 99, 106, 117, 120, 129, 141, 143.)

Finchampstead (666). A village three miles south-west of Wokingham. (pp. 96, 98.)

Hagbourne, East and **West** (1360). Villages near Didcot junction, both very attractive, with old cottages and half-timbered houses. There are two village crosses and part of a third. In the church at East Hagbourne are good examples of Transition Norman and of all the later styles of architecture. The chancel arch is Transition, the tower arch and chancel Early English, and the tower Perpendicular in style. (p. 86.)

East Hagbourne Village

Hampstead Marshall (244). A village three and a half miles south-west of Newbury. There is a beautiful deer park, the house in which was burnt in 1718 and has not been rebuilt. (p. 137.)

Hampstead Norris (760). A village and railway station on the Didcot-Newbury line and in the chalk district. The church has a Norman doorway and an Early English chancel, and the staircase to the rood-loft remains. (pp. 21, 90, 138.)

Hendred, East and **West** (1038) are villages between Wantage and Didcot, both most attractive, with half-timbered houses and churches of mixed styles but with many points of note. Hendred House with an old chapel attached is of considerable interest. (p. 116.)

Hungerford (2364). A market town on the old Roman road to Bath on the river Kennet, a part of the town being in Wiltshire. It is a great resort of anglers. Charles I was at the

Hurley Church and Site of Lady Place

Bear Inn, November 1644, and at the same inn William of Orange met the commissioners from James II in 1688. (pp. 8, 13, 18, 22, 80, 119, 120, 142.)

Hurley (493). An interesting village on the Thames with old houses, four miles north-west of Maidenhead. (p. 105.)

Hurst (1214). A village three miles north-west of Wokingham. (p. 101.)

Ilsley, East (482).   A small town in the chalk district two and a half miles from Compton station with a large sheep market. The Duke of Cumberland, uncle of George III, had a house and training stables here, and it is now a great place for training horses.   The church is mainly Early English.   (pp. 80, 118.)

Inkpen (658).   A village four miles south-east of Hungerford. To the south of the village there is a range of chalk hills, the highest of which is Inkpen Beacon, 975 feet above the sea. Walbury Camp is a large earthwork on the same range a little to the east, with an altitude of 959 feet.   (p. 12.)

Kintbury (1548).   A large village with a railway station nearly midway between Newbury and Hungerford.   It is on the river Kennet.   Brick-making is carried on in the neighbourhood and there is a whitening factory.   The church is largely Norman with an Early English tower.   (pp. 60, 61.)

Lambourn (1476).   A small town in the midst of the chalk district with a light railway to Newbury (12 miles).   It is an important centre for training race-horses.   The river Lambourn is a good trout stream.   There is an old market cross.   The large church is cruciform with a central tower which is Norman in character with small round-headed windows.   Much of the church is Transition Norman.   The east window is of the Perpendicular period.   (pp. 85, 143.)

Lockinge, East (301).   A village two miles south-east of Wantage.   The church, mainly of the Decorated style but with a good Norman doorway, has been recently enlarged.   Lockinge House stands in a beautiful park close to the village.   (p. 91.)

Maidenhead (10,757).   A municipal borough and market-town on the Thames with a railway station 24½ miles from Paddington.   The borough is partly in Bray and partly in Cookham parish.   There are grain mills and breweries, and some trade in timber is carried on.   (pp. 12, 17, 20, 75, 90, 114, 119, 120, 123, 138.)

Marcham (798). A village two and a half miles west of Abingdon with many stone quarries in the neighbourhood. (pp. 34, 89.)

Mortimer. See **Stratfield Mortimer.**

Newbury (8924). A municipal borough and market-town with a railway station on the Great Western 53 miles from Paddington, and also with railways to Didcot, Southampton, and Lambourn. The borough was incorporated by charter of

Pangbourne

Elizabeth. The chief trade is in agricultural produce. There are maltings and corn mills. The town has large new municipal buildings, a free library, a district hospital, and a large grammar school as well as many charities. A race-course has recently been made a little to the east of the town with a separate railway station. (pp. 17, 22, 59, 70, 76, 79, 86, 89, 101, 111, 115, 119, 128, 129, 137, 145.)

Pangbourne (1235). A village with a railway station 41½ miles from Paddington, situated at the junction of the river Pang with the Thames. (pp. 10, 21, 85.)

Radley (444). A village with a railway station 58 miles from Paddington. Radley College, a large public school, is situated a mile to the west of the village. (pp. 133, 138.)

Reading (52,660). A county, municipal, and parliamentary borough, and the county town of Berkshire. It is a most important railway centre 36 miles from Paddington and is served by the South Western and South Eastern as well as by the Great Western railways. It has, in fact, excellent railway communication with every part of England and Wales. The charter of incorporation was granted by Henry III.

Reading is situated on the river Kennet close to its junction with the Thames. There are large municipal buildings with a free library and an excellent museum, a county hospital, a university college, a grammar school, and many other schools and charitable institutions. The Berkshire County Hall and the Assize Courts are at Reading and are close to the old gateway of Reading Abbey. The few remains of the abbey are now the property of the Corporation and are laid out as gardens adjoining the public Forbury garden. The railway works are extensive and there are iron foundries, engine and agricultural implement works, cycle works, electric-light works, printing works, a very large establishment for making biscuits, and also one for the production and sale of seeds. There are also flour mills, breweries, brick and tile works, steam launch and boat-building yards, and establishments for making ropes and sacks. St Mary's church is said to have been built of materials from the ruins of the abbey. The walls are largely of a chequer pattern of dressed flints and squares of freestone. (pp. 6, 19, 22, 36, 54, 58, 62, 64—84, 89, 90, 99, 103, 104, 108, 125—133, 140, 144.)

Sandhurst (2386). A village on the river Blackwater four and a half miles south-east of Wokingham with a railway station on the South Eastern and Chatham railway. The Royal Military College is two miles south-east of the village near Blackwater station. (pp. 130, 132.)

Shefford, Great or **West Shefford** (422). A village between Lambourn and Newbury. The church has been already mentioned. (p. 117.)

Shottesbrook Church from the Park

Shinfield (1015). A large village three miles south of Reading.

Shottesbrook. A park four miles south-west of Maidenhead. The beautiful church has been already mentioned. (pp. 63, 108.)

Shrivenham (951). A village with a railway station on the Great Western 71½ miles from Paddington, near the border of Wiltshire. It gives its name to the Hundred. (pp. 72, 138, 143.)

Sonning (526). A very attractive village on the Thames two and a half miles below Reading, the parish is partly in Oxfordshire. In the tenth and eleventh centuries there was a Bishop of Berks and Wilts and the palace was at this place. The church is large with Early English arches and many monuments.

Sparsholt (646). A village three and a quarter miles west of Wantage. There is a fine church in the Decorated style. (p. 99.)

Streatley Mill

Stanford in the Vale (853). A village nearly four miles south-east of Faringdon. The church, in mixed styles, is interesting. The tower is Early English, there is a squint from the north aisle to the altar, and a very curious piscina with a reliquary above it.

Steventon (797). A village with railway station on the Great Western three and a half miles south-south-west of Abingdon. There is a raised flood-path by the road through the village, a number of old houses, and a church in mixed styles with a south aisle and tower arches of the Decorated period. (pp. 101 108, 117.)

Stratfield Mortimer (1405). A village and residential district with a railway station named Mortimer on the Reading and Basingstoke line, the nearest station to the Roman town of Silchester in Hampshire.

Streatley (562). A village on the Thames opposite Goring in Oxfordshire, with which it is connected by a bridge. This is a very old crossing place and the Ridgeway is directed towards this point. (pp. 19, 90, 118.)

Sunningdale (1409), five miles south of Windsor, with a station on the London and South Western, was until recently a district of heath and pine woods, but it is being rapidly built over and good golf links attract many visitors. (p. 86.)

Sunninghill (2479). A village and residential district close to the above. Two chalybeate springs, Sunninghill Wells, were a fashionable resort in the eighteenth century. (pp. 106, 118, 135.)

Sunningwell (289). A village two miles north of Abingdon. Bishop Jewel was vicar and is said to have built the singular octagonal porch at the west end of the church.

Sutton Courtney (1295). A village on the Thames two miles south of Abingdon. The abbey, the manor house, and the manor farm were buildings connected with Abingdon Abbey, and are all of interest, dating from the twelfth to the thirteenth century. In the church the chancel arch and walls are Transition Norman, the tower arch is Norman. (p. 114.)

Swallowfield (1375). A village on the river Blackwater five miles south of Reading. The church has a wooden bell-cot with very fine old timber work. A Bible of 1613 is preserved in the church and Miss Mitford's grave is in the churchyard. (p. 144.)

Thatcham (2177). A large village three miles east of Newbury which was once a small town with a market. There is some good Norman work in the church.

Three Mile Cross. (p. 144.)

Tidmarsh. (pp. 21, 30.)

Tilehurst (5965). A village on the plateau two miles west of Reading with a considerable brickmaking industry.

Wallingford Bridge

Twyford (1106). A small town in Hurst parish four miles north-east of Reading, with a railway station on the Great Western, the junction for the Henley line. (pp. 17, 142.)

Uffington (518). A village in the Vale of White Horse about six miles west of Wantage with a railway station 66¼ miles from Paddington, the junction for the Faringdon line. There is a large cruciform church mainly dating from the Early English period. The central tower is octagonal. Uffington Castle is a large earthwork on the chalk downs close to the White Horse and two miles south of the village. (pp. 5, 7, 86, 88, 98.)

Upton (338). A village with railway station on the Didcot and Newbury line, two and a half miles south of the former and on the edge of the chalk district. The church is a small Norman chapel of early character.

Wallingford (3049). A municipal borough and market-town 51 miles by rail from Paddington and 46 miles by road from London. It is situated on the Thames and is built on a wide area of river gravel. The charter of its incorporation dates from the time of Henry II. There is a bridge over the river built in 1809 on the site of an older structure. The town hall with an under-croft of 1670 has been already mentioned (page 128). There is a corn exchange, free library, and grammar school. The trade is in agricultural produce and malt. On the three sides of the town away from the river are very ancient earth ramparts, and the keep-mound and some slight remains of a Norman castle still exist. (pp. 57, 59, 65, 67, 68, 70, 71, 73, 74, 80, 85, 88, 91, 106, 111, 128, 142, 143.)

Waltham St Lawrence (867). A village four and a half miles south-west of Maidenhead. (pp. 62, 90.)

Waltham, White. See **White Waltham.**

Wantage (4146). A market-town in the Vale of White Horse. The railway station, Wantage Road, is nearly two and a half miles from the town. There are ironworks but otherwise the trade is mainly in agricultural produce. The church is large, cruciform and in mixed styles. Wantage was the birthplace of Alfred the Great, and Butler, the author of the *Analogy*, was also a native. (pp. 30, 67, 119, 122, 137, 140.)

Warfield (919). A village in Windsor Forest with an interesting church mainly in the Decorated style. A mile to the north-west is the steeplechase course of Hawthorn Hill. (p. 99.)

Wargrave (1857). A large village on the river Thames between Reading and Henley. (pp. 20, 23, 144.)

White Waltham (679). A village three miles south-west of Maidenhead. Prince Arthur, son of Henry VII, lived in the manor house, now a farm. (pp. 64, 142.)

Wickham. A village on a clayey hill five and a half miles north-west of Newbury. The very old church tower has been already noticed. (pp. 95, 119.)

The Stocks at White Waltham

Windsor or **New Windsor** (13,958). A municipal borough and market-town 22 miles from London. It is a parliamentary borough, a large part of which is in Buckinghamshire. The town, which has grown up round the castle, was incorporated by Edward I. The High Steward is H.R.H. Prince Christian. There is a town hall, public library, and reading room, and both cavalry and infantry barracks. Windsor Castle has long been a favourite residence of our Kings and Queens. (pp. 20, 62, 68—80, 89, 95, 99, 104, 108, 114, 117, 128, 134, 142.)

Windsor, Old (1962). A village two miles south-east of the castle. It was the residence of Edward the Confessor. The church is a small one in the Early English style. Beaumont College is in this parish.

Winkfield (1026). A village in Windsor Forest four and a half miles south-west of Windsor. Foliejon Park is a little to the north of the village. (p. 35.)

Wittenham, Long (470). A village on the Thames between Abingdon and Wallingford with an interesting church of mixed styles but mainly of the Decorated period. (pp. 91, 101.)

Wokingham (5923). A municipal borough and market-town with a railway station 36¼ miles from Waterloo. The charter of incorporation was granted by Queen Elizabeth. There is a town hall and also a number of charitable endowments, one of which was founded by Archbishop Laud. The trade is mainly in agricultural produce, timber, bricks and tiles. (pp. 60, 128, 134, 140.)

Wytham (230). A village in the most northern corner of Berkshire, close to Oxford. The church is built of material which was mostly brought from Cumnor Hall. Wytham Abbey, a building of the sixteenth century, is close to the church and has a fine park. (pp. 114, 139, 140.)

Yattendon (274) stands on a clayey hill five and a half miles west of Pangbourn. The church, built about 1450, is a good example of the Perpendicular style. There are some extensive and ancient underground galleries in the chalk near this place. (pp. 85, 139.)

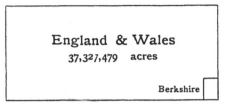

Fig. 1.   Area of Berkshire (462,208 acres) compared with
          that of England and Wales

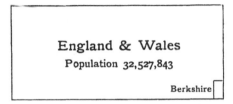

Fig. 2.   Population of Berkshire (256,509) compared with
          that of England and Wales in 1901

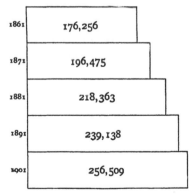

Fig. 3.   Increase of population in Berkshire from
          1861 to 1901

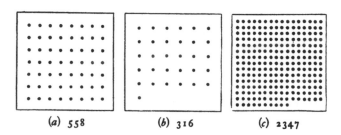

(a) 558   (b) 316   (c) 2347

Fig. 4. Comparative Density of Population to the square mile
in (a) England and Wales, (b) Berkshire, (c) Lancashire

(*Each dot represents* 10 *persons*)

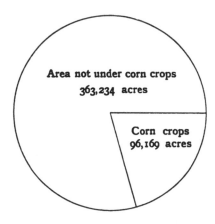

Fig. 5. Proportionate Area under Corn Crops in
Berkshire in 1908

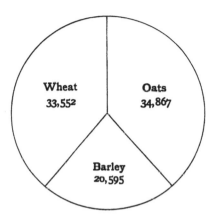

Fig. 6. Proportionate Area in Acres of chief Cereals
in Berkshire in 1908

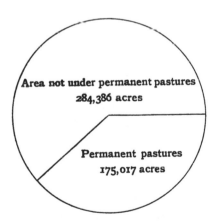

Fig. 7. Proportion of Permanent Pasture in
Berkshire in 1908

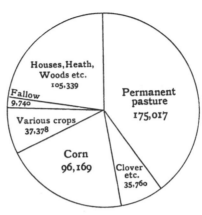

Fig. 8. Proportion of Permanent Pasture to other Areas in Berkshire in 1908

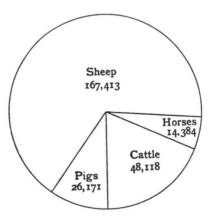

Fig. 9. Proportionate numbers of Live Stock in Berkshire in 1908

Milton Keynes UK
Ingram Content Group UK Ltd.
UKHW041520181024
449640UK00009B/88